The Synthetic Age

The Synthetic Age

Outdesigning Evolution, Resurrecting Species, and Reengineering Our World

Christopher J. Preston

The MIT Press
Cambridge, Massachusetts
London, England

This book was set in Stone Serif by Jen Jackowitz. Printed and bound in the United States of America.

Library of Congress Cataloging-in-Publication Data

Names: Preston, Christopher J. (Christopher James), 1968- author.
Title: The synthetic age : outdesigning evolution, resurrecting species, and reengineering our world / Christopher J. Preston.
Description: Cambridge, MA : MIT Press, [2018] | Includes bibliographical references and index.
Identifiers: LCCN 2017029324 | ISBN 9780262037617 (hardcover : alk. paper)
Subjects: LCSH: Technology--Social aspects. | Technological innovations.
Classification: LCC T14.5 .P75 2018 | DDC 303.48/3--dc23 LC record available at https://lccn.loc.gov/2017029324

10 9 8 7 6 5 4 3 2 1

For Toby, Jessica, and Alice ... whose lives will be shaped by the Synthetic Age

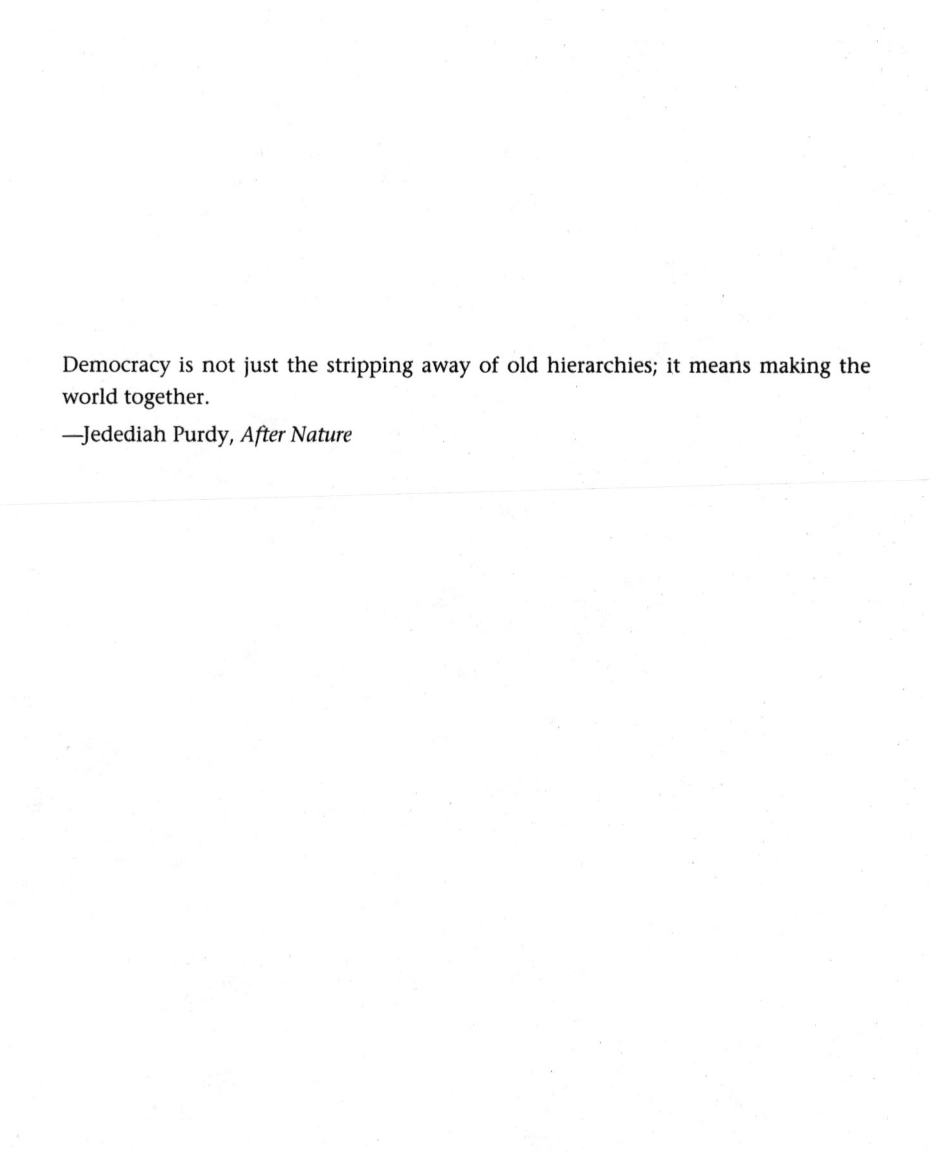

Democracy is not just the stripping away of old hierarchies; it means making the world together.

—Jedediah Purdy, *After Nature*

Contents

Acknowledgments

My wife, my parents, my brother, and my sister are tireless cheerleaders for who I am and for what I do. I appreciate from the core of my being all that they have given me over many years.

A number of friends, colleagues, and acquaintances have played roles in keeping this manuscript buoyant at various points on its voyage. Among others, I am grateful to Fern Wickson, Svein Anders Lie, Geoff Gilbert, Jake Hanson, Patrick Kelly, Armond Duwell, Neal Anderson, Jennifer Beck, Jack Rowan, Beth Clevenger, Ted Catton, and Bradley Layton for their supply of both information and encouragement.

Finally, my agent Kevin O'Connor was of more assistance than I could possibly have deserved in my efforts to navigate the literary world. Provocative, hard-working, informative, funny, and dedicated, it is because of him that you now hold this book in your hands. Kevin proved himself to be, without doubt, one of the most capable guides a writer could hope for.

Introduction

Whoever you are—a scientist or painter, a farmer or philosopher, a young mother or wrinkled grandparent—a radical shift in how you look at the world typically begins with a single moment of awakening. In one instant, something happens that crystallizes a whole set of thoughts and observations into a shocking new realization. Such a moment happened for me not long ago off a remote Alaskan coastline in the company of a grizzled fishing boat captain named Walt.

• • •

It was 2 p.m., and I was perched on the back deck of a forty-two-foot boat with a nasty-looking gaff[1] in my hand, watching a quarter mile of fishing line emerge from the sea.

"You ready?" Walt demanded. "When a fish comes up, you need to be quick."

I nodded and shuffled my feet to make sure they were gripping the deck firmly, hoping not to screw up my first attempt at landing an Alaskan halibut destined for the commercial market.

"You don't want to lean over too far," Walt added, "or one of the big suckers will pull you in. They fight like hell when they come to the surface."

I signaled my understanding and tightened my fingers on the boat's rail. The halibut in the waters off Alaska can weigh twice as much as a man and can cause havoc on a small boat. Some fishermen put a bullet in the halibut's brain before hauling it aboard to avoid risking injury when the fish starts thrashing around on deck.

With my heart thudding in my chest, I looked down to where the dripping line was emerging from the sea, just in time to see a huge oval shape sweep into view.

Nine hours after the silhouette of that first fish appeared alongside our boat, we pulled into a remote cove in the shadow of Mount Fairweather. Below deck, the fish hold was packed with a thousand pounds of our quarry, their cleaned-out bellies stuffed full of shaved ice. As we glided into the cove, a brown bear on the beach looked up from a salmon gripped between two giant paws and then quickly returned to his meal. After the anchor was set and the captain killed the noisy diesel engine, only the water lapping against our hull and a few shrieks from passing gulls broke the heavy liquid silence.

It was nearly midnight, and I was exhausted from working all afternoon with the heavy fish. But in the northern twilight, I sat for a long moment on the back deck in my sweaty fishing gear and took in the mountains, the glaciers, and the fading outline of the bear on the beach. Mentally and physically worn down by the work, a sad realization washed over me: I finally grasped what it meant to say that humans have utterly transformed the earth.

Besides our boat, there were no signs of people in any direction. These beautifully sculpted fish had been taken from some of the most remote coastal waters in North America, waters teeming with species in numbers found in few other places. If there was anywhere left on earth where some semblance of pristine nature could still be found, it would be in a place like this.

Yet the glistening white flesh of the halibut we had pulled from the ocean, cleaned meticulously with our knives, and stacked in ice below decks was not pristine. It contained enough mercury spewed from Chinese coal-fired power plants four thousand miles away that U.S. Food and Drug Administration numbers suggested a safe consumption limit of only three small portions a month. Pregnant women and small children should eat even less.

As someone whose regular job involved teaching college students about environmental issues, I already knew in the abstract that there were no longer any places left on earth untouched by industrial pollution. Although this information had lodged somewhere in my brain, I clearly had not fully processed it. Because now, for the first time, I really *felt* it. Human impact on the planet means more than just a sequence of numbers pointing toward sagging snow packs, melting glaciers, and shrinking species counts. It means a landscape that can no longer shrug off the consequences of human industry, no matter how far from the manufacturing and urban centers you go. The human stamp on the world is total. Nor are these impacts trivial. Even in far-away places, this human imprint can affect the safety of the food we put into our mouths.

In the months since I returned from that fishing trip, I have wondered what this legacy means for the times ahead. The question this book seeks to investigate is "Where are we heading from here?"

• • •

Until recently, virtually all of the notable pieces of human history have taken place in an epoch known as the Holocene. Derived from the Greek terms *holos* and *kainos*, Holocene literally means "entirely recent." The planet has occupied the "entirely recent" epoch for a geologically brief twelve thousand years or so.

Over the last decade, a varied collection of climate scientists, ecologists, and geographers has been suggesting that humanity's outsized influence on the earth means we are on the cusp of leaving the Holocene behind. This chastening new reality is frequently now referred to as the arrival of the Anthropocene or the "human age."[2] Technically speaking, the *Anthropocene* is a geological term, one that—if you want to speak *really* technically—does not actually refer to anything yet. It is the new name under consideration for the geological epoch that will replace the Holocene. A growing cohort of commentators have suggested that the coming epoch should be named in honor of the species whose signature is now detectable on every square inch of soil and in every drop of ocean water.

Despite its felicitous sound, *Anthropocene* is not the only term being used to capture this shifting moment in the earth's history. Other words for the emerging epoch have been suggested, each reflecting a different conception of what a human-dominated planet really means. Some have proposed the terms *Capitalocene* or *Econocene* in order to capture something about the role played by business in the transition that the planet is experiencing. Others think the word *Homogenocene* would better characterize the diminishing of human and biological diversity on display. Some feminists think the term *Manthropocene* more suitably speaks to the question of which portion of humanity has wrought the bulk of the planetary havoc. A parallel line of thinking has proposed the term *Eurocene*, and more downbeat voices have suggested simply the *Obscene*.

More important than what we choose to name this new period of geological history, however, is how we choose to *shape* it. The emergence of a new epoch is not simply a chance to rename a planet we have unwittingly transformed through our labor and industry. It is an opportunity to think carefully about the world we will choose to create. And on this score, we live in a remarkable time. At the very moment this naming discussion is taking place, a new age is dawning. From the atom to the atmosphere, a suite of technologies is emerging that together promise to remake the natural world.

• • •

In the 1967 movie *The Graduate*, the dazed-looking hero, Benjamin Braddock (played by Dustin Hoffman), is pulled aside by a well-meaning family friend and told that the key to his future can be found in a single word: "Plastics." As the friend saw it, an awful lot of the stuff Ben would see around him was going to be synthesized in factories using new types of cheap and highly flexible chemical processes. If Ben knew what was good for him and his career, he needed to make himself a part of it.

Today, if Ben were getting such advice, he would hear a much grander promise of an even more startling synthetic future. Humans are no longer just surrounding ourselves with new materials. Our species also is gaining the ability to reengineer a number of key planetary processes.

We are learning how to synthesize and stitch together new arrangements of DNA to build original and useful organisms. We are fabricating novel atomic and molecular structures to create entirely new material properties. We are reassembling the species composition of ecosystems, while experimenting with bringing extinct animals back from the dead. We are studying how to deploy technologies that will turn back the sun to keep the planet cool. In each of these ways, humanity is learning how to replace some of nature's most historically influential operations with synthetic ones of our own design.

Nobody would deny that many major planetary transformations have already occurred. Up to this point, however, most of the major global impacts our species have wrought have all been inadvertent. Nobody planned to sully Alaskan coves with mercury or allow industrial chemicals to penetrate the flesh of whales swimming beneath Arctic ice. Neither the atmospheric warming attributable to the burning of fossil fuels nor the mass extinctions from widespread habitat destruction were deliberate. In all of the transformations to date, global change has been far from the minds of the perpetrators.

From now on, however, things will be different. After we fully awaken to the global nature of the damages we have inflicted, we have no option but to make our decisions about future actions more self-aware. Like the injured animal we find suffering by the side of the road, the broken planet has suddenly become our responsibility. We no longer have the option of turning away and pretending we have not noticed. Good conscience will no longer permit it.

To make matters worse, the responsibility is now particularly acute. At the very time we must assume this moral burden, new technologies are making possible an even deeper transformation of the surrounding world than anything that has occurred before. A number of the earth's most basic functions—how DNA is constructed, how sunlight penetrates the atmosphere, how ecosystems are composed—can increasingly be determined by human design. What used to be the unplanned result of natural processes is now more and more a product of our conscious decisions. When discussing the future we will inhabit, Nobel laureate in chemistry Paul Crutzen gives blunt expression to what lies ahead. From now on, he says, "It is we who decide what nature is and what it will be."[3]

The replacement of natural processes with synthetic ones is the hallmark of what might be called a Plastocene epoch. This term is not chosen to suggest a world full of plastic. Humanity may find reasons to move away from this particular synthetic creation over the coming decades. The term *Plastocene* reflects the adjectival use of the word *plastic* and indicates a planet that is becoming increasingly pliant and moldable. The Plastocene speaks to the unprecedented degree of malleability of the Earth that new technologies are making possible for those with the resources to develop and deploy them.

By deliberately tinkering with some of the planet's most basic physical and biological operations, humans stand on the verge of turning a world that is found into a world that is made. In the Plastocene, the world is thoroughly reconstructed, from the ground up, by molecular biologists and engineers, marking the beginning of the planet's first Synthetic Age.[4]

The remaking of the planet during this Synthetic Age will not be just a matter of changing surfaces. It will reach deeply into the earth's metabolism. The technologies driving this new epoch will change not just how the planet *looks* but also how the planet *works*. Nature and the processes that run it will increasingly become something we design.

Understanding the character of these transformations is important because crucial choices need to be made. The exact contours of the path ahead are not yet fixed. We need to decide how far into remaking the earth we should go. Although some level of management of natural processes is now inevitable, the Plastocene could still take many different forms depending on how aggressively we choose to impose our designs.

According to one approach, the new relationship to the earth required in the coming epoch will finally reject the idea of stepping back and making an effort to diminish our footprint on the planet. Instead, it will quickly ramp up human intervention into nature and its processes. Rather than impacting nature thoughtlessly and accidentally, a "full-throttle" Plastocene means that we would shape it confidently, deliberately, and sometimes ruthlessly, all according to the best abilities of our technical experts. Nothing would be off-limits.

Others balk at this high level of intervention and see the dawning of the new epoch as an opportunity to dial back our meddling. Even as we intensify the management of nature in some areas, we could become increasingly less involved in others. By choosing to treat certain stretches of DNA as inviolable, for example, we could ensure the protection of some portion of what evolution handed down to us. By designating some landscapes as entirely off-limits, we could preserve some important symbols of the earth's wildness and independence. At the same time that we encourage the development of certain planetary-scale technologies for humanitarian reasons, we could push back against other aspects of an increasingly synthetic world.

With many of the questions about the shape of this Synthetic Age still unanswered, we occupy a crucial transitional moment, a fleeting opportunity for reflection as the planet enters a different period of its history. At the very time we are finally recognizing the extent of our impacts, my suggestion in the pages that follow is that the debate about what sort of future we desire needs to stay open for a little while longer. Rather than assume that the epoch ahead already has our species' name stamped all over it, let us assume that we occupy a brief but important thinking space. Invoking Janus, the Roman god of transitions with one face for looking backward and another for looking forward, this moment provides a window of opportunity to survey the accidental impacts of the past and to carefully consider the deliberate impacts of the future.

The recent wave of populism in European and U.S. politics has been interpreted to mean that more and more people fear that control of their future is slipping away from them. Their life, it appears to them, is increasingly in other people's hands. If we fail to behave thoughtfully in this transitional moment, the contours of the Synthetic Age will indeed be shaped by distant experts and by economic interests. The decisions about how much to remake the earth will be made by technical elites and by the marketplace, each of them lured by some combination of genuine altruism and the prospect of new profits toward ever more drastic interventions. In such a case, if we let ourselves be dragged by commercial interests thoughtlessly into a full-throttle Plastocene, a momentous shift will be thrust upon us. The earth and many of its fundamental processes will lose their independence from us. In some real and final sense, our surroundings will be robbed of their naturalness. The biosphere will become entirely subsumed under the technosphere.

Such happenings will have consequences. In doing this to the earth, we will ultimately be doing something to ourselves.

• • •

Let me be clear from the outset that this book is not a rejection of the important areas of research and discovery described here.[5] Starting at the level of atoms and moving up to manipulations of the whole atmosphere, the chapters ahead celebrate a number of the powerful technologies currently emerging. There is no doubt that many of these developments will be necessary to cope with the impacts that are being created by an increasingly urbanized and industrialized population. These technologies will allow more humans to live better lives with less impact than ever before. Some of these tools also will be essential for repairing the damages that already have been done. To a great extent, some version of the Synthetic Age is inevitable.

The inevitability of some of these transformations comes with a sober warning, however. Within the promises of the technologies lurk some dangerous seductions. They often involve exaggerated fantasies about control. They put us in a planetary manager's role for which we are little prepared. And they dissolve a long-standing pact about the way humans should aspire to treat the world that surrounds them.

The remaking of ourselves and the earth offered by the Synthetic Age presents a distinctly double-edged sword. There certainly will be many benefits gained. But there also will be significant costs. On occasion, it will mean a joyous new vision of health and affluence and an optimistic exploration of new types of relationships with our surroundings. At other times, it will create a desperate fight to cling to our sanity in a world rapidly becoming unrecognizable from the one we inhabited in the past. We will find ourselves running quickly and blindly across uncertain and uneven terrain.

The future we will inhabit is guaranteed to be different, but the shape it will take is yet to be determined. In a just world, this shape would be decided by careful and informed popular choice. This is one of the central messages I hope to convey in what follows. These are not decisions that can be left in the hands of a select few. After all, the stakes for our species could hardly be higher.

Key Figures

Name	Dates	Short description
Diane Ackerman	b. 1948	Author of one of the first popular books on the Anthropocene, *The Human Age: The World Shaped by Us* (2015)
Jennifer Beck	b. 1973	U.S. Park Service botanist at Crater Lake National Park and advocate of proactive restoration techniques for the threatened whitebark pine
Paul Bogard	b. 1966	Author of *The End of Night: Searching for Natural Darkness in an Age of Artificial Light* (2013)
Stewart Brand	b. 1938	Futurist and former environmental entrepreneur; pursuing deextinction of species such as the passenger pigeon through the Long Now Foundation
Francis Collins	b. 1950	Government scientist and former head of the Human Genome Project; director of the U.S. National Institutes for Health
Michael Crichton	1942–2008	American science fiction author; created a public backlash against nanotechnology with his novel about runaway nanobots titled *Prey* (2002)
Paul Crutzen	b. 1933	Nobel Prize–winning atmospheric chemist from the Netherlands; widely credited (with Eugene Stoermer) with introducing the idea of the Anthropocene to a broad audience; first major public figure to endorse research into climate engineering; leading advocate of aggressive intervention into natural systems

Name	Dates	Short description
Eric Drexler	b. 1955	Futurist and cofounder of The Foresight Institute; a pioneer in molecular manufacturing in nanotechnology; blamed for his role in creating nanotechnology's "grey goo" public relations disaster
Richard Feynman	1918–1988	Nobel Prize–winning physicist, author, musician, and public servant; helped launch the field of nanotechnology with a lecture in 1959; served on the Rogers Commission that investigated the explosion of the space shuttle *Challenger*
Stephen Gardiner	b. 1967	Expert in climate ethics and author of *A Perfect Moral Storm: The Ethical Tragedy of Climate Change* (2011); notably cautious about climate engineering
Jay Keasling	b. 1964	Leading synthetic biologist and creator of a process to make semisynthetic artemisinic acid, a precursor to an important antimalarial drug
David Keith	b. 1964	Harvard professor and energy policy expert; strong advocate for research into climate engineering
Ray Kurzweil	b. 1948	Nanotechnologist, futurist, and expert in artificial intelligence; inventor of the keyboard synthesizer and text-to-voice technologies; author of *The Singularity Is Near: When Humans Transcend Biology* (2005); advocate of transhumanism
Keekok Lee	b. 1938	Philosopher of technology and the environment responsible for developing a critique of "deep technologies"
Aldo Leopold	1887–1948	Early American conservationist and author of *A Sand County Almanac* (1949); known for attributing high moral value to wild and relatively pristine landscapes
Jason Mark	b. 1975	Journalist and editor of *Sierra Magazine*; cofounder of Alemany Farm; author of *Satellites in the High Country: Searching for the Wild in the Age of Man* (2015); an advocate for the contemporary significance of wildness

Name	Dates	Short description
Emma Marris	b. 1979	Science writer and author of *Rambunctious Garden: Saving Nature in a Post-Wild World* (2011); a leading proponent of the new, interventionist type of environmental thinking
Bill McKibben	b. 1960	American climate activist; his book *The End of Nature* (1989) provided an important early alert about the philosophical significance of climate change; a strong advocate of more restraint and respect for nature's independence from us
John Stuart Mill	1806–1873	Nineteenth-century British political philosopher and reformer whose essay "On Nature" (1874) set up a dichotomy between the idea of human action existing entirely within, or entirely outside of, nature
Svante Pääbo	b. 1955	Swedish genomicist famous for his work on mapping the genome of the Neanderthal
Fred Pearce	b. 1951	Freelance British science journalist and author of the book *The New Wild: How Invasive Species Will Save the World* (2015)
Richard Smalley	1943–2005	Nobel Prize–winning chemist and nanotechnology pioneer; known for his codiscovery of the Buckminsterfullerene
Chris Thomas	b. 1959	Biologist and expert on the ecological effects of climate change; pioneered the technique of assisted migration by moving two butterfly species northward by car in the UK
J. Craig Venter	b. 1946	Synthetic biologist whose intervention into the Human Genome Project dramatically brought forward the date by which the genome was sequenced; led the group that in 2010 succeeded in building the world's first living synthetic genome
Gaia Vince	Not available	Australian travel and science writer and author of *Adventures in the Anthropocene: A Journey to the Heart of the Planet We Made* (2014)
Sergey Zimov	b. 1955	Russian ecologist and director of the "Pleistocene Park" in Siberia

1 Making New Matter

Benjamin Franklin, Karl Marx, and Hannah Arendt are among the respected historical figures who have suggested that *Homo sapiens* ("wise hominid") might better be known as *Homo faber* ("building" or "tool-making hominid"). Our penchant for constructing things—from pyramids to shopping malls and battery-powered Teslas—is one of the primary activities we perform. Arguably, it is the essential thing that makes us who we are. The desire to build objects and devices seems to be written into our DNA. The fact that we cannot stop ourselves from doing so has been the key to our spectacular success as a species relative to all the other feathered and furry fauna that roam the planet.

Even though there are literally millions of artifacts one can buy from yard sales, street markets, shops, and tacky websites across the globe, nature has always placed limitations on our construction projects. Certain properties of the material world have set limits on the things that can be built. You cannot make a furnace out of a tub of water, for instance, and you cannot construct a functioning airplane from a stack of bologna sandwiches. Despite the ingenuity and skill that humans have put into making things, the nature of matter has always determined certain boundaries or limits. However much you bend, cut, mix, cool, or forge some material, there are certain things that it simply will not become.

Or so it has seemed. The advent of nanotechnology has suggested an upending of this fundamental truth.

American theoretical physicist Richard Feynman is widely credited with giving birth to the field of nanotechnology. It happened during a remarkable lecture delivered in 1959 at the California Institute of Technology. We will get to what he said below, but first it is important to know something about the man who gave this ground-breaking speech.

The term *renaissance man*—applied to an individual so broadly talented that he or she can impart wisdom or conjure amazement on just about any topic—probably understates the character of Richard Feynman. He was primarily a leading theoretical physicist and mathematician. But Feynman was also an accomplished bongo player, a best-selling author, a translator of Mayan texts, a part-time artist who sketched under the pseudonym of "Ofey" (adapted, said Feynman, from the French *au fait*, meaning "it's done"), and a renowned storyteller who possessed a wicked sense of humor that he frequently wielded to great effect.

A Nobel Prize–winner in physics (1965), Feynman also is remembered as a distinguished national servant. As a young man, after some initial hesitation, he was part of the team at Los Alamos, New Mexico, that developed the nuclear bomb that helped end World War II. In the final years of Feynman's life, President Ronald Reagan asked him to serve on the commission investigating the fatal explosion of the space shuttle *Challenger* in 1986. In a televised public hearing into the disaster, which killed seven astronauts, Feynman dropped a clamped rubber O-ring into a cup of iced water to show how the temperature at the launch site would have interfered with the proper elastic behavior of the seals on the *Challenger*'s fuel tank. In this simple way, Feynman effectively demonstrated the cause of the explosion to the watching American public. Although he was then suffering from a terminal form of gastric cancer, Feynman looked long and hard at the assumptions and biases that had shaped the space shuttle program as a whole. He calculated that the chance of a catastrophic disaster on any one shuttle mission was not 1 in 100,000, which engineers at the National Aeronautics and Space Administration (NASA) had always publicly suggested, but closer to 1 in 100 (a statistic tragically borne out during the shuttle fleet's thirty-year service life).

Part of Feynman's brilliance lay in the fact that he was always skeptical of institutional mindsets and the overconfidence they fostered.

During his time at Los Alamos, he was so concerned about the possibility that the nuclear technology they were developing would get into the wrong hands that he taught himself to be an expert safe-cracker. His supervisors laughed, but shortly after World War II ended, Feynman broke into the safe containing all the files necessary to build the bomb, thus proving his point about institutional complacency to his supervisors. In practice and in theory, Feynman knew how to shed light on problems lurking right under the noses of those around him.

In that Cal Tech lecture hall in 1959, Feynman's topic was something much more theoretical than O-rings and Cold War secrets. Addressing some of the brightest physicists in America, Feynman speculated about what things actually looked like down at the scale of atoms and molecules. At the time, everybody suspected that this scale brought one close to certain absolute physical limits where the nature of things was pretty much fixed. Yet in this talk, called "There Is Plenty of Room at the Bottom," Feynman hypothesized that there was actually sufficient space available deep inside any individual piece of matter for humans to start rearranging and manipulating the particles they would find there. Through a somewhat mind-bending discussion of the space available on the head of a pin, the amount of print in the *Encyclopedia Britannica*, and the quantity of information stored in DNA, Feynman portrayed the atomic scale as an environment with enormous potential for manipulation. Such intentional reconfigurations, he proposed, would create the possibility of making extraordinary stuff happen. It was, he claimed, a research area ripe for exploitation.

In that pioneering lecture, Feynman predicted that atoms and molecules would one day be directly manipulated, using specially designed tools to create new materials with staggeringly useful properties. He confidently stated that when humans gained control of the arrangement of atoms, they would discover "an enormously greater range of possible properties that substances can have, and of different things that we can do."[1]

The speech was remarkably prescient. In 1959, scanning tunnel microscopes able to "see" at the atomic scale did not yet exist.[2] So nobody could actually confirm whether Feynman was right. Nevertheless, Feynman's predictions launched scientists and engineers on a revolutionary new path toward remaking the physical world.

• • •

The nanotech revolution began stealthily. The first consumer products containing nanomaterials entered commercial markets in 1999. Well before the public had any clue about what nanotechnology was, car bumpers coated with paints containing nanomaterials that could resist scratches, tennis racket frames embedded with carbon nanotubes for strength, and sunscreens with nanosized reflective agents to repel ultraviolet light started appearing in stores. Consumers started buying them and incorporating them into their daily lives. The extraordinary physics involved in nanomaterials remained hidden from the unassuming customer.

The prefix *nano* stands for 10^{-9}, as in one billionth. The many zeros in this prefix suggest that a billionth is a pretty small fraction of anything. When the metric is meters, this fraction translates into a very small length. Things measured in nanometers are really very tiny indeed. One nanometer is about one hundred thousandth of the thickness of a sheet of paper. There are more than 25 million of them in an inch. A strand of DNA buried deep within the nucleus of a single cell of your body is already two nanometers in diameter. If a glass marble could be reduced to the size of a nanometer and everything else was shrunk down proportionately, an average adult could step over the earth in a single stride (provided, of course, that the adult had not also been shrunk down).

If you prefer a different reference to the body, fingernails grow by roughly one nanometer every second. Even if you stare at those nails really, really hard, you cannot see them lengthening. By contrast, Ryan Gosling's—or pretty much any movie star's—stubbled beard grows by five nanometers per second (and people certainly do keep staring).

The startling nature of nano continues. A water molecule is less than half a nanometer long. A gold atom is even smaller (closer to a quarter of a nanometer). A typical bacterium, on the other hand, is a massive 2,500 nanometers wide, while basketball player LeBron James is an epic 2.03 billion nanometers tall.

This raises an important point. Things tend to stop being considered nanoscale after they exceed one hundred nanometers. Beyond that, they become *macro*. This means that neither the bacterium nor LeBron James is nano. On the other hand, as long as a material is nanosized in at least one dimension, it counts as nano. Graphene, for example, is a lattice of carbon that is never more than one atom thick. A sheet of graphene with the diameter of a dinner plate counts as nano because the graphene "plate" is no more than a nanometer from top to bottom. The illustrations all point in the same direction. A nanometer is very small, and nanoscience is the study of the properties of matter at these very small dimensions.

Although the study of the nanoscale is a relatively new area of scientific research, a number of free-floating nanoscale things have been present on the earth since long before *Homo faber* got into its fabricating ways. Scattered nanosized entities can be found in soils, in ocean waters, and in the atmosphere. Some of nature's most captivating phenomena—such as the sheen on a butterfly's wing, the stickiness of a gecko's feet, or the slipperiness of the rim of a carnivorous pitcher plant—rely on nanosized biological structures present in each organism. Unusual nano carbon structures like graphenes and fullerenes—which basically are balled-up graphenes—occur naturally not only on the earth but also in space.

Humans also have inadvertently created nanomaterials on occasion. Centuries-old stained glass owes some of its beauty to the presence of nanosized gold and silver particles, although the artisans creating the glass had no clue that they were utilizing the nanoscale. Damascus swords more than a thousand years old have been found to contain individual carbon fullerenes on their blades. The quality of the aromas that circle a fresh-brewed pot of coffee or the offensive odors that emanate from a festering pile of wet garbage rely on properties present at the nanoscale.

Despite the occasional presence of a few nanosized materials in the natural environment (and their inadvertent and sporadic production by humans over the centuries), the vast majority of materials and elements exist in nature at scales thousands of times greater than the nanoscale. Why the nanoscale became so interesting to scientists after Feynman has a lot to with the reason for the rarity of nanomaterials.

At the nanoscale, materials tend to be highly reactive and promisingly unstable. This means that, left alone in nature, they usually will quickly react with nearby substances to become something bigger and more inert. Nanotechnology has become one of the hottest areas in science and engineering precisely because researchers have figured out how to make materials so that they can exploit this intense reactivity at the nanoscale *before* it has had the chance to react and to become more boring and stable.

With intentionally manufactured nanomaterials, the ordinary quickly becomes the extraordinary. Nanosized flour can explode when exposed to a flame, gold can change its color to red and have the temperature at which it melts plummet, and unlike other forms of the element, carbon in the nanoform conducts electricity very well. Nanodots glow in strange but controllable ways when illuminated with light, materials can be made orders of magnitude harder by giving them nanosurfaces, and nanosubstances can be used to catalyze intense chemical reactions. In the strange world of nano, supermagnetic properties can suddenly emerge, and the direction of the magnetic field can flip randomly under the influence of temperature. Across a range of domains, the act of shrinking materials creates a wholly new and exciting reality.

There are some elementary physical truths that underpin this wizardry that are quite illuminating and do not require a lab-coated PhD's grasp of theoretical physics. The source of a nanomaterial's intense reactivity and unusual properties is in large part a matter of basic geometry. If you take any sphere and shrink it in size, then the ratio of its surface area to its volume goes up. This means that a very small piece of material has less on the inside relative to how much it has on the outside. Thus, a small marble has a bigger surface-area-to-volume ratio than a big marble. A really tiny marble has an even bigger surface-area-to-volume ratio than the small one.

A consequence of this large surface-area-to-volume ratio is that a much greater proportion of the material is found at the surface and exposed to the outside world. Chemical reactions between substances happen at the surface. So with all that surface area exposed, a larger portion of the stuff in question is on hand to be involved in reactions. These reactions make possible numerous interesting things.

As materials shrink progressively down in size toward the nano range, the surface area to volume ratio starts to become ridiculously large. For example, a particle ten nanometers across has 20 percent of its atoms at the surface. A particle three nanometers across has about 50 percent of its atoms at the surface. That is a lot of stuff on the outside! With all this exposed surface area, it is not surprising that materials develop chemical and physical properties that the same substance does not possess at larger scales.

Geometry, however, is not the entire story. Another reason properties change dramatically at the nanoscale has more to do with matter itself. At larger scales, the rather spooky effects present in the quantum world are rarely noticeable because they are averaged out over the millions of atoms that make up the whole material. Because there are considerably fewer atoms involved with any material at the nanoscale, the averaging out of quantum properties that is always occurring at larger scales is no longer so normalizing. Quantum effects can therefore start to dictate the material's behavior.

Think of it this way. If ten thousand people yell obscenities at you, it is likely you will hear only a noisy and indistinct roar. If only five or six people yell the same obscenities, you are likely to hear enough to be offended. Something similar happens at the nanoscale, where a handful of quantum properties can start to really get heard.

Quantum effects occur in part as a consequence of the discrete energy bands in which electrons vibrate inside a material. When the size of a material is shrunk down until it approaches the size of these bands, the behavior of the electrons changes. These changes can significantly influence a material's optical, mechanical, thermal, magnetic, and electrical properties, adding extra spice to the surface area effects. Carbon nanotubes—which look a bit like nanoscale penne pasta—conduct heat very well from one end of the "penne" to the other, but they insulate highly effectively across the tube. Graphene is typically nonmagnetic, but it can become magnetic after briefly being wrapped in certain materials.

Both graphenes and nanotubes, thanks to their nano dimensions, also have the unusual optical property of being superabsorptive of light. This makes them into some of the blackest materials available, which is helpful for their use with laser technologies. Nanotubes also hold together very well, possessing several times the tensile strength of steel at a fraction of its weight. This phenomenal strength is vastly different from macro forms of carbon. Graphite, as you may recall from countless hours spent breaking pencil tips as a schoolchild, is pretty fragile. Nanoscale graphene, by contrast, is a suitable material for bulletproof vests.

The powerful properties of matter that scientists can now hijack at the nanoscale clearly contain enormous potential. If cheap and common materials like carbon can suddenly become lighter, stronger, more flexible, more conductive, and more magnetic simply by being manufactured at a different size, then whole fields of endeavor gain new and exciting possibilities. These fields include materials science, health care, information technology, energy production, optics and sensoring, military technology, and commercial manufacturing. The list goes on. On contemplating the possibilities, the Nobel laureate and nanotech pioneer Richard Smalley declared giddily, "The list of things you could do with such a technology reads much like the Christmas wish list of our civilization."[3] Whatever you want, you can have. Nanotechnology has potential application in almost any domain that *Homo faber* fabricates.

The novelty of the properties emerging across the nanoscale is something that an important subspecies of *Homo faber*—*Homo faber economicus*—immediately realized contained enormous economic potential. If humans could invigorate matter by resizing it to expose its most unusual and valuable properties, a whole world of promises opens up. Within these promises, a vast amount of money could be made. This is part of the reason the U.S. government now invests around $1.5 billion annually into the National Nanotechnology Initiative, a broad effort to promote invention and discovery at the nanoscale across the U.S. economy.

• • •

Nearly two decades into the modern nanotechnology revolution, it is hard to keep track of the many areas in which nanomaterials are influencing commerce. Nanotreatments that modify surface behaviors make numerous household items more water-repellent, antireflective, ultraviolet-filtering, antifogging, and antimicrobial. Golf clubs, sunglasses, window coverings, food supplements, kitchen appliances, and children's toys all contain nanomaterials. Nanocoated fabrics can resist red wine and ketchup spills. Silver nanoparticles embedded in the armpits of shirts make people stink less by killing the bacteria responsible for body odor. Food packaging that includes nanosilver can resist harmful microbes and increase shelf life. Nanostructures embedded in packaging also can better seal in desirable features such as the carbonation in fizzy drinks. Nanotreated fridges and freezers stay cleaner. Nanoparticles in cosmetic products perform functions that range from increasing the penetration of the product into the skin to enhancing the evenness of lotion's application. Cutting tools with blades incorporating nanosized materials can be many times more durable than their non-nano counterparts.

Nanotechnology already has proved its mettle in the information technology arena at the user interface. Smartphone screens using nanostructured polymers produce sharper images with less glare. Bendable screens promise mobile devices that can be put in your back pocket and sat on without causing an expensive trip to the iStore. But these developments are only the superficial ones. Even more potential lies in how the nanoscale promises to speed up the processing of digital information.

As Feynman theorized in his talk, small size suggests incredible possibilities for information storage and handling. In today's computers, these functions are performed by transistors made out of semiconducting materials such as silicon. The transistor typically contains two terminals, between which a current can be switched on or off by applying a voltage to a third terminal known as the gate. As these transistors have shrunk, the tiny distances between the terminals have meant that the technology not only is fiendishly expensive but also is approaching the point at which a strange phenomenon called quantum tunneling occurs.

Quantum tunneling has the unfortunate consequence of allowing electrons to flow inadvertently between the gate and the channel separating the other two terminals, even when this space is insulated. Due to this and other electron leakage, unwanted heat is generated, efficiency is reduced, and the zeros and ones that are needed to represent digital information are no longer guaranteed.

One potential answer to this problem is to replace conventional transistors with transistors made out of nanowires. Due to the structure and dimensions of these nanowire channels, the current through them can be reliably controlled with minimal leakage of electrons. A more radical approach gets rid of transistors entirely, instead utilizing the binary nature of the spin of atoms and electrons. Researchers are beginning to learn how to flip these spins virtually instantaneously. A third option being explored by Dutch scientists uses the positions of atoms to capture ones and zeroes. These researchers have figured out how to store information at a density that is two or three orders of magnitude beyond current technology by moving individual chlorine atoms into different positions on a copper plate.

Astonishing computing power is possible with data processing at this scale. Information processors could be much smaller and more energy efficient than anything available today. This additional computing power would allow the development of user-friendly functions impossible today, including the ability to almost instantaneously store data during a system crash.

From opening the door to highly efficient methods of data processing to providing the utterly mundane convenience of a stain-resistant carrier bag, nanotechnology is proving itself to be a transformational technology across numerous domains of modern life.

• • •

In the midst of all the excitement generated by its tremendous potential, let's pause a second to reflect on just what a radical move nanotechnology represents. Nanotechnology takes some of the fundamental parameters of the material world that humans evolved into and recalibrates them. The standard forms of matter presented to us by the earth can now be significantly restructured. By shrinking things down to the nanoscale, fabricated nanomaterials can provide new building blocks with new characteristics that nature herself had mostly concealed. Veils that hid useful behaviors are lifted. These *new* forms of *old* types of matter can serve us in ways that earlier incarnations of *Homo faber* could not possibly have imagined. By entering the nanoscale, humans pry the lid off a world that until now history shielded from our gaze, a world almost entirely unknown and unutilized by previous generations.

Nanotechnology promises a level of intervention into nature that is more profound than anything preceding it, and in so doing the technology subtly recalibrates the relationship between humanity and the physical stuff of the world. We do not need to be content with the existing forms and properties of the materials we find or even with the standard structures of the elements we have identified. Nanotechnology allows us to uncover new properties by tweaking existing atomic and molecular arrangements. The material limits of familiar forms of matter no longer apply. Nanotechnology effectively makes available to us a whole new dimension of the material world.

Plenty of environmentalists have decidedly mixed feelings about the idea of manipulating matter at the atomic and molecular levels. To some, it feels like a step too far. It seems as though there is a reason that the unusual properties revealed at the nanoscale have remained mostly hidden from view. The intense reactivity exposed is unfamiliar and alarming. The fact that in the normal course of things these unusual properties are unavailable to us says something important. Investigating the nanoworld can feel to some like poking a slumbering serpent that might better be left alone.

Although this hesitancy is understandable, ecologically conscious doubters have to concede that nanotechnology could make massive contributions to environmental sustainability. In the realm of energy, nanostructures designed for their thermoelectric properties can capture waste heat from wherever it is leaking and turn it back into electricity. Developments in nanotechnology are already contributing to more efficient solar technologies that can feed more powerful and more quickly rechargeable batteries. Nanotechnology creates the possibility of flexible or even paintable photovoltaic panels that could be daubed on anything that sits in the sun—from your car to your garage door to your dog.

Used as catalysts, nanomaterials can make combustion more efficient and help break down woody plant materials for quicker conversion into biofuels. Special optical properties enable nanoparticles to be used as indicators of the presence of environmental contaminants. Highly reactive nanotreatments can help pull these contaminants out of dirt or water and remediate sites saturated with difficult-to-extract pollutants. A type of gold nanostructure is being developed that can pull carbon dioxide out of the atmosphere using only solar power. Graphene sheets acting as nanofilters may be able to sift hydrogen out of the air in the way that a net sifts salmon from ocean waters. The hydrogen could then be burned as a clean fuel that leaves water as its only by-product.

Another environment for which nanotechnology might prove helpful is the human body. Just as nanotechnology promises great environmental benefits, it also is starting to have a formidable range of applications in healthcare. The quantum properties displayed by certain nanocrystals provide huge advantages for medical imaging in the body, including a longer fluorescence of injected materials over a wider spectrum of light than anything that has been available in the past.

When inserted into the body, these so-called quantum dots tend to interfere less with the behavior of any cellular material the diagnostician is attempting to scrutinize.

Nanosensors are being designed that will be able to detect molecular changes in cells. This creates the possibility of spotting malignancies far in advance of what current technologies permit. Nanomaterials also have been shown to be capable of spurring the growth of optical and spinal nerves, offering the possibility of enhanced recovery from debilitating injuries. Nanostructures are already playing a role in bone and tooth implants by providing better surfaces for improved integration of prosthetic materials with the patient's jawbone. Nanosized fat particles laced with toxic drugs can be delivered to tumors and then, on excitation with a gentle heat, can be made to release the drug in the desired location without harming neighboring cells. These "thermal nanogrenades" currently under development have been tagged by medical experts as "the holy grail of nanomedicine."

The wish list that appears to be on offer at the nanoscale sets the minds of engineers and inventors into a creative spin. Want to deliver a payload cheaply into space? How about a space elevator? Such a device would use a very long cable stretching between the ground and an orbiting platform to zip a payload from the earth's surface to beyond the reaches of gravity. Concerned about creating a cable strong enough but light enough to stretch that far? No problem. Weave it out of carbon nanotubes. Nanotechnology makes such impossible visions possible. Space elevators are merely—if a space elevator can be "merely" anything—one of the most futuristic tips of a simply massive nanotechnology iceberg.

• • •

When you add up everything its boosters promise and survey the whole package with a cool and deliberate eye, you might start to wonder whether nanotechnology sounds a little bit too good to be true. There is no doubt something hugely exciting about the possibilities on offer. But as is often the case with a powerful emerging technology that its enthusiasts promise will transform our lives immeasurably for the better, *Homo faber*'s powerful nanotech blade has the potential to cut both ways.

There is good reason to be hesitant about the idea of novel and highly reactive forms of matter being deliberately introduced into many sectors of our daily lives, sectors that include our food, our clothing, and our bodies themselves. Nanotechnology has economic potential precisely because the properties on display are new. This means that, for the most part, our species did not evolve alongside these materials, and it is unclear what their enduring consequences might be, both for us and for the surrounding environment. Although people might disagree about whether to define nanomaterials as "unnatural"—after all, nanosubstances have always been lurking in small quantities across the natural world—humans are not used to encountering them frequently and intimately in our daily lives.

In a campaign that has parallels to the debate about genetically modified organisms, some consumer advocates think there is enough uncertainty about the effects of these novel structures on human and environmental health that there should be clear labeling of commercial products that incorporate nanomaterials. In most countries, such labeling is currently not required. As a response to this lack of information, various online inventories are trying hard to keep up with the stream of new nano products entering the market.

One of the most comprehensive of these lists was developed by the Project on Emerging Nanotechnologies (PEN) in Washington, D.C. Because the development of consumer products containing nanomaterials has been happening very quickly, the PEN list no longer claims to be comprehensive. Nevertheless, it currently contains close to two thousand products available for purchase that are believed to contain some form of nanomaterial.

The list's authors relied on the product manufacturers to identify the presence of a nanomaterial. Sometimes the manufacturer's claim is broadcast on the material's packaging as a selling point. Other times, perhaps anticipating a potentially negative public reception, the presence of nanomaterials is kept relatively quiet. Different countries displayed different sensitivities to the publicity around nanomaterials. Banana Boat, for example, was worried about consumer concerns over its sunscreens and released a statement in Australia in 2012 declaring, "No nanoparticles (i.e. particles smaller than 100 nanometres in size) are used in any Banana Boat sunscreens manufactured and sold in Australia."[4] However, the company's U.S. webpage is silent on the issue. In many cases, the compilers of the PEN database admit, it is not possible to corroborate independently a manufacturer's claim.

The PEN database includes information about potential routes of exposure to nanomaterials for an individual using that product. These routes include the skin, the lungs, and the stomach because various nanoproducts are designed to be held, inhaled, or eaten. A recent study revealed that gold nanoparticles inhaled into the lungs can travel around the body in the bloodstream where they can lodge themselves in various vulnerable locations with uncertain consequences for the vascular system.[5] Responding to consumer concern, European law now dictates that cosmetics, foods, and nutritional supplements containing nanoproducts sold in the European Union must be labeled as such. There is no similar requirement in the United States. For the most part, nanosized materials enjoy the same regulatory regimes in America as the macrosized materials from which they are drawn. The U.S. regulatory system has until now assumed that an atomic structure is an atomic structure, whether it is confronted at the nanoscale or in larger forms. This rationale sidesteps the fact that it is precisely the differences in properties found at the nanoscale that make nanomaterials interesting.

New reporting requirements for nanomaterials adopted by the U.S. Environmental Protection Agency (EPA) in January 2017 for the first time required manufacturers and processors of nanomaterials—although not the companies ultimately *selling* the nano product to consumers—to provide some basic information about what they are manufacturing and processing. The stated purpose of the rule is to assist the EPA in assessing whether any further regulation of nanomaterials is required and to create an inventory of what exactly is currently being manufactured. In a reassuring gesture toward business interests, the final rule emphasized that nothing it requires is based on the assumption that "nanoscale materials as a class, or specific uses of nanoscale materials, necessarily give rise to or are likely to cause harm to people or the environment."[6] Offering still more reassurance later in the text when discussing other federal statutes that might be relevant, the rule states that the new requirement would serve as a useful tracking measure but "does not concern an environmental health or safety risk."

In truth, the newness of the technology and the lack of conclusive, long-term research conducted so far means that the health and environmental consequences of nanomaterials in many cases remain uncertain. The territory is complex, and there is an interesting conundrum buried in the debate over how to regulate nanomaterials that dogs many of the technologies of the Plastocene. It concerns whether the distinction between the natural and the artificial can still be a reliable guide at the dawning of the Synthetic Age. Traditionally, there has been a tendency to associate the natural with the *normal*, the *ecological*, and the *safe*. The synthetic or the artificial, on the other hand, has been associated with the *humanized*, the *unnatural*, and (often) the *potentially suspicious*. The safety of synthetic products has always been a legitimate subject for scrutiny.

This gross generalization has never been a dependable one. Many naturally occurring substances (such as arsenic or snake venom) are deadly, while many artificial products (such as synthetic insulin or neonatal intensive care units) can be literally life-saving. But as a broad rule of thumb, this generalization has retained its popularity because it appears to lean on some deeply held cultural assumptions that there is something profoundly reassuring about what is natural. The labels on products that line the shelves of health food stores testify to this fact. These assumptions have grown in strength alongside the modern environmental movement.

Nanomaterials can really mess with these conventional rules. They are in many cases derived from common substances, large numbers of which are regarded as perfectly safe. The ones that are not are already regulated in the United States under the Toxic Substances Control Act of 1976. According to the Nanotechnology Industries Association, 85 percent by weight of all the nanomaterials currently produced are derived from carbon or silicon, neither of which are elements that ring many alarm bells for high toxicity. But this is surely no reassurance. If the nanoworld did not generate startling new properties, it would not be of much interest to researchers and business. And if these startling properties have been exceedingly rare through human history, our bodies are unlikely to be adapted to them.

Commercial interests simply cannot have it both ways. If a particular form of a material displays behavior that is dramatically different from the norm, it probably ought to be given a little extra scrutiny.

History is littered with technological promises that turned out to serve a vested minority very well while paying scant attention to the unwitting public on which they were foisted. These cautionary tales deserve to be kept in mind during this early phase of the nanotechnology revolution. New forms of matter rubbed onto our skin, entering our lungs, and passing through our colons present risks. It would be foolish not to study these risks before welcoming these materials into so many facets of our lives.

Nanomaterials are just one embodiment of a recurring dilemma in this new Synthetic Age. Obviously, there are huge potential benefits for health and environment to be gained from a technology that can so radically rearrange the world around us. Equally obvious, this level of manipulation of our surroundings creates a reason for caution.

Alongside these practical concerns about health and safety, the more philosophical point should not escape notice. With the advent of nanotechnology, something about our relationship to the world around us shifts. Nanotechnology allows our species to insert itself into the very nature of matter in a way that humanity has not done before. It attempts an unprecedented rearrangement of the materials that nature provides. Nanotechnology is not only novel *empirically* in the sense of producing forms of matter that are largely new to science. It is also novel *conceptually* in the sense that it takes humans further into the business of reshaping the world than our species has ever ventured before. At stake are not just questions of risk and benefit but also deep questions of meaning and value. Embarking on a nanotechnological future demands asking just how far into the natural order humanity should probe.

Such a technology needs to be considered carefully not only by research scientists and risk assessors but also by philosophers, futurists, wizened elders, and purveyors of traditional knowledge from around the world. These are not merely commercial decisions. They are important decisions about who we want to be. For this reason, justice demands that they be as richly democratic as possible. This would appear to be one of the most basic demands of the coming Synthetic Age.

2 Repositioned Atoms

New material behaviors made available at the nanoscale turn out to be just one part of the nanotechnology dream. Feynman's 1959 lecture contained an additional vision for nanotechnology that went far beyond simply uncovering more useful material properties. He predicted a powerful human future that utilized specially developed tools to place atoms and molecules directly into carefully worked-out arrangements. The far-sighted physicist saw that if one could rearrange atoms by picking them up and moving them about, it should be possible to build pretty much anything you wanted, one atom at a time . Atoms could be used as the most basic of all construction materials. He called this nanotech vision *molecular manufacturing*.

In 1989, three decades after the Cal Tech speech, the first step toward Feynman's dream was accomplished. Researchers at IBM showed it was feasible to pick up individual atoms manually and place them in a new location. They used the tip of a scanning tunneling microscope to move thirty-five individual xenon atoms into an arrangement that spelled out the letters of their employer: I-B-M.[1] These researchers showed that, with the right tool, it was possible to reposition atoms into newly chosen configurations.

The ability to place individual atoms in the arrangements of your choice is part of what initially intrigued Feynman about the nanoscale. By controlling where every atom goes, you theoretically could gain the potential to construct anything you could imagine. You also could do it with very little waste. Any collection of material elements could be reconstituted to construct pretty much anything else. The sheer number of atoms available gives a hint of the potential. A bucket of water is thought to contain more hydrogen and oxygen atoms than the Atlantic Ocean contains buckets of water. A pile of household garbage would contain trillions of atoms from a wide range of existing elements all potentially available for repositioning.

Molecular manufacturing therefore would offer unlimited potential for repurposing. If you could separate out the atomic components contained within a big stack of bologna sandwiches and knew how to rearrange them, then perhaps one day you really could make a functioning airplane. Atoms of the particular elements are similar to each other whether they are found in carbon fiber airplane wings or in processed meat sandwiches. This repurposing signals not only a rethinking of what counts as waste but also a vastly different sense of what counts as material limits.

The idea of molecular manufacturing once again sends creative minds into a frenzy. Rather than focusing on the recycling and repurposing materials, the goal of molecular manufacturing championed by Feynman was to move atoms around to construct nanoscale robots that would perform useful tasks. These tiny machines—known as nanobots, nanoids, or nanites—could be given assignments that would be unimaginable for macroscale devices.

In his 1959 address, Feynman referenced how he and a colleague had mused about how it would be "interesting in surgery if you could swallow the surgeon." In a vision later brought to cinematic life in the 1966 movie *Fantastic Voyage*, they imagined nanosized minisubmarines moving autonomously through arteries—perhaps eating a bit of plaque here, inspecting some discoloration there—until they arrived at the heart. On arrival, they could directly examine the chambers, report back to real surgeons monitoring on the outside, and make critical interventions. Fleets of them could perform life-saving cholesterol clean-ups. Nanobots could be designed to find and destroy lymph-born cancer cells or to obliterate harmful viruses. Future nanoscale robotic surgeons might be small enough to operate on individual neurons, creating the potential to make injured people walk or even think again.

Outside of the medical arena, advocates envision numerous valuable roles for these miniscule roving workhorses to play. Nanites could eat smog or clean up spilled chemicals. They could seek out and destroy bacteria in drinking water. Due to their small size, nanites could venture to places that humans cannot go without being detected. In a military arena, they might conduct espionage missions in hostile environments or take defensive action in the skies against incoming chemical threats.

Although Feynman was particularly fond of them, the idea of well-designed molecular machines doing valuable work for us is not limited to the creation of miniature free-ranging robots. YouTube is filled with animations of hypothetical production facilities containing spinning ratchets, rods, propellers, and cassettes that look much like what you might find within any contemporary factory until you notice they are picking up and depositing atoms and molecules instead of bits of wood, metal, or plastic. These perfectly functioning nanomachines would spend their days layering out atoms and molecules into desirable arrangements. Rather like a 3-D printer but on a much smaller scale, the machines represented in these animations build things layer by layer—literally, atom by atom. It suggests an infinitely more productive and less wasteful future in which precision forms of labor take place at the nanoscale rather than the macroscale.

It is not surprising that the vision of molecular manufacturing is so appealing. The benefits of miniaturization are already familiar to us. Doing things smaller has paid well-known dividends in consumer product areas such as electronics and information storage. The idea of doing something with "atomic precision" has virtually become a synonym for doing something as effectively as it can be done. This makes nanorobotics appear to be the solution to almost any problem that sounds like it could be tackled by the repetition of small and precisely coordinated mechanical operations.

The enthusiasm, however, needs tempering. Although first proposed by Feynman nearly sixty years ago, the reality of nanobots and molecular manufacturing still lies far in the future. Most of the action so far has been on the material properties science side of nanotech. The

National Nanotechnology Initiative that began in 2000 under President Bill Clinton devotes relatively little money to the speculative promise of molecular manufacturing, instead using most of its resources in areas that are already proving their commercial potential.

In the meantime, research into future nanomachines has taken a surprising turn. The current state of the art of molecular manufacturing looks less like building robots using the principles of mechanical engineering and more like building biological structures using the principles of biochemistry. Researchers in molecular manufacturing quickly realized that the best examples of nanoscale machines performing useful functions at the atomic and molecular scales were the biological "machines" found in the cells of living organisms. The cutting edge of today's molecular manufacturing research involves trying to recreate nature's own molecular nanobots by building in the laboratory biological structures that can perform simplified versions of the actions that the real things do in the bodies of organisms.

By carefully mimicking what they see going on in molecular biology, researchers have constructed biobased molecular "motors" that rotate when light is sent in their direction. They have designed molecules that can "walk" along a prescribed track and molecular "cranks" and "ratchets" based on protein structures that can be made to rotate and move items along paths. They have even built a device, dubiously labeled a *nanocar*, that has four rotating fullerenes, located on "axles," that looks vaguely like a wheeled vehicle and lurches in one direction when excited by a jolt of energy. All of these advances are in a field colloquially known as *wet* or *biomimetic* nanotechnology, a name chosen because it copies the water-based operation of the structures found in the living organisms studied by molecular biologists.

Despite some interesting successes, it turns out that nature is a good deal better at wet nanotechnology than humans are. One contemporary survey of the state of molecular manufacturing concluded that even though molecular machines are the basis of every significant biological process, "none of mankind's fantastic myriad of present-day technologies exploit controlled molecular-level motion in any way at all."[2]

Efforts to do so have had only limited success. For example, autonomous motion of a human-designed molecular machine in the presence of a "fuel" has been difficult to reproduce. The machines that have been built have only a single functioning part. Researchers also have been unable thus far to ensure long-term stability in the biological machines they have created. As a result, a whole set of unanswered questions about what molecular manufacturing is really trying to achieve has arisen. The researchers in the field have been split on the question of whether they should be copying the "machines" found in the bodies of organisms or using biological components to copy the inorganic machines humans already build at the macroscale.

In other words, progress in molecular manufacturing to date has been halting and perhaps a little underwhelming. The seventy-eight-page study of molecular manufacturing's progress also noted that manufacturing in biological systems always takes place in solution, something that adds several layers of inconvenience if you want to do future molecular manufacturing in a dry factory rather than in a biological environment. Despite the rather discouraging conclusions found throughout the analysis, the authors suggest that the future for molecular manufacturing remains "bright." Optimism in the face of long odds is part of what keeps science moving.

Molecular manufacturing's problems, however, do not end with its desperately slow progress. The entire field also fights with a considerable public image problem. The whole idea of molecular manufacturing is hindered by a debilitating and self-inflicted public black eye. Ironically, the black eye inadvertently was caused by a person who had been one of the technology's foremost advocates.

• • •

From the start, Eric Drexler was precocious and visionary. At the age of twenty-six, two decades after Feynman's famous speech, he published a paper in the *Proceedings of the National Academy of Sciences* that detailed the mechanical principles behind deliberately rearranging atoms and molecules. Appearing at the start of the Reagan era, a period of can-do national optimism, "Molecular Engineering: An Approach to the Development of General Capabilities for Molecular Manipulation" abruptly rekindled the fire that Feynman had lit. Five years before he got his doctoral degree, Drexler followed this paper with *Engines of Creation: The Coming Era of Nanotechnology*, a book-length exploration of the potential for nanobots that succeeded in creating a whole generation of nanotech dreamers virtually overnight. In the same year, with his former wife, Christine Peterson, Drexler founded the Foresight Institute, which promotes cutting-edge nanotechnology for the development of "transformative future technologies" in pursuit of the public interest.

When Drexler finally turned his attention to getting his PhD in molecular nanotechnology from the Massachusetts Institute of Technology in 1991, it was the first PhD of its kind anywhere in the world. But Drexler had never hesitated about charting his own path. His pursuit of Feynman's molecular manufacturing vision had been single-minded, pioneering, and relentless, driven forward by an unquenchable optimism.

Two and a half decades later, little had changed. His book *Radical Abundance: How a Revolution in Nanotechnology Will Change Civilization* (2013) suggests a physical world that can be remade by humans so dramatically that material limits virtually disappear. In it, he continues to advocate for the world-changing possibilities presented by taking advantage of all that "room at the bottom." The Feynman Institute—no longer headed by Drexler—continued to operate out of its Palo Alto headquarters in California, issuing a range of Feynman Prizes to those most effectively implementing the vision of molecular manufacturing. But although the vision lived on, a good deal of the optimism about molecular manufacturing Drexler generated early in his career ended up being drained by a major public relations disaster that was entirely of his own making.

In *Engines of Creation*, Drexler includes a chapter titled "Engines of Destruction." In this sixteen-page essay toward the end of the book, he expresses his concern that manufactured nanobots and molecular assemblers could become highly destructive if they were designed with the ability to feed and to reproduce themselves—so destructive that they could obliterate the entire planet.

Self-replication and self-fueling are highly desirable if one wants to create molecular machines that perform useful functions on a meaningful scale. Anything manufactured at the nanoscale is by definition extremely small. In most cases, nanobots would have to be created in vast numbers in order to be of any practical use at a human scale. The clean-up of a polluted industrial site or the manufacture of a valuable material for a city's infrastructure would require literally trillions of nanites to work at any practical scale. The best way to gain these numbers would be for the nanobots to have the capacity to create more of themselves to generate sufficient a volume of "workers" to perform the task. Furthermore, to maintain their labors, they would have to be powered by sunlight or by some other ambient energy source that each one of them could acquire or ingest on its own.

Self-replicating and self-fueling nanobots are all well and good in terms of efficiency, but they come with a distinctly dark side. Every time these increasing numbers of microscopic workers feed or replicate themselves, they have to grab material from the world around them to use as fuel or source material. Drexler understood this completely. In a casual moment in which he probably intended to do no more than explain the science he found so fascinating, molecular manufacturing's self-appointed guru pointed out that this self-replication and continual consumption had the potential to spin out of control.

Anyone who has been bored to tears listening to a statistician or economist talk about the power of compound interest growth has a sense of how quickly self-replicators might take over. Populations of anything that double within a repeatable length of time become frighteningly big frighteningly fast. Drexler pointed out that self-replicating nanobots able to duplicate themselves at the rather unambitious rate of once every one thousand seconds would make over 68 billion copies of themselves in a mere ten hours.

Like the piranhas in a low-budget sci-fi movie, these relentlessly more numerous nanobots could go on a feeding frenzy that ended up consuming everything around them. The source material they would need to live and reproduce would lead to a devouring of everything in their path. The impact of an exponentially increasing population of nanobots would be catastrophic. These self-nourishing and mobile machines would, Drexler warned, "reduce the biosphere to dust in a matter of days." "Replicators," cautioned Drexler in his famous chapter, "give nuclear war some company as a potential cause of extinction." Even worse, unlike nuclear war, out-of-control nanobots would not be hard to produce. "To devastate Earth with bombs would require masses of exotic hardware and rare isotopes," Drexler suggested as he dug himself a bigger and bigger hole. "But to destroy all life with replicators would require only a single speck made of ordinary elements."[3]

Futurists dubbed the phenomenon of ravenous, world-consuming nanobots "runaway global ecophagy," a term that not only sounded disgusting but also pointed toward utter catastrophe for all involved. Those who were already somewhat inclined to be skeptical about the whole idea of nanotechnology were alarmed to find that one of the world's most enthusiastic nanotech advocates was himself worried about the possibility of turning the earth into an undifferentiated mass of "grey goo." From whichever corner of the Plastocene you look at it, this was not going to be a pretty fate.

It did not take long for Drexler to appreciate the firestorm he had unleashed. His nightmare about uncontrolled nanobots wreaking havoc quickly became the subject of several works of science fiction. Michael Crichton's novel *Prey* (2002), a sensationalizing of the destructive potential of nanobots, reached number one on the *New York Times* bestseller list and was made into a Hollywood movie. Drexler's speculations became a major public relations disaster for the whole idea of molecular nanotechnology and the nanobot. Even Prince Charles became concerned, calling on Britain's Royal Society to investigate nanotechnology's threats to the Crown.

Drexler realized how unhelpful to the cause his grey goo warning had become and tried to tamp down the fuss. He took the unusual step of coauthoring an academic paper that attempted to dismiss his own idea. In "Safe Exponential Manufacturing," Drexler and coauthor Chris Phoenix reasoned that such out-of-control nanobots would run out of energy or turn cannibal before destroying the earth. Nanites might also not need to be self-replicating in the first place. Besides, others chimed in, rather than letting them get out of control, you could take a few sensible precautionary steps such as what amounted to screwing your nanofactory and its nanobots down to the floor to preventing anything from running riot.

After a number of years of intensive damage control, the panic instigated by the fear of global ecophagy began to wane. Most researchers in the nanotech community do not even think about the dangers of self-replicating nanobots anymore. They have better things to do with their time and their research dollars. Only the more cautious—or perhaps mischievous—of them still acknowledge that nothing about the possibility of runaway global ecophagy appears to contravene any of the known laws of physics.

As a result of the grey goo episode, Drexler's swashbuckling image was severely tarnished. He found himself being passed over for grants and advisory positions. His version of nanotechnology fell out of favor and was replaced by the simpler—if less inspiring—vision of uncovering interesting material properties at the nanoscale.

For the first time in his career, Drexler felt like he was on the outside of a revolution he had believed he was leading. To make matters worse, the fallen hero soon found himself engaged in different but perhaps more personally damaging fight. This was a fight not with Hollywood or the technology's public image but with a fellow nanoscience pioneer—the same one who had previously suggested that nanotech offered a "Christmas wish list" for civilization. The fight was about the whole coherence of molecular manufacturing from a theoretical point of view. Unhappily for Drexler, his adversary this time was a scientist for whom he held a great deal of respect and admiration.

• • •

Richard Smalley was very much the opposite of Drexler in habit and temperament. Born in Akron, Ohio, Smalley followed a relatively traditional, if highly distinguished, academic path. He completed a PhD in chemistry at Princeton University, and then did a postdoctoral fellowship at the University of Chicago before taking up a job at Rice University, where he spent the rest of his research career. Rather than being a precocious visionary who published books and founded think tanks before he was out of graduate school, Smalley embarked quietly on years of painstaking work surrounded by successive cohorts of loyal PhD students and postdocs in his lab on Rice's campus in Houston.

The work he did was brilliant. In 1996, Smalley shared the Nobel Prize in Chemistry for discovering the Buckminsterfullerene, an unusual spherical form of carbon that looks like a soccer ball. He discovered it accidentally, so the story goes, while trying to simulate the atmospheric conditions surrounding the formation of stars. From this success, his lab work took him deeper and deeper into the curious chemistry that attended the formation of nanostructures. For most of his career, Smalley was industrious, introverted, and reclusive. Only as his fame spread did he venture reluctantly into the public eye. In the last decade of his life, he started to use his clout as a Nobel Prize winner to speak about what he saw as the world's most significant upcoming challenges, including renewable energy production, the provision of clean water, and global public health. The battle with Drexler, however, was about the fundamentals of science.

Smalley thought that Drexler's mechanistic vision of molecular manufacturing and nanobots showed a failure to understand how atoms and molecules worked in the real world. Atoms and molecules, Smalley insisted, are not like Lego pieces that you can physically place wherever you want and snap together at will according to your designs. They are constrained by the properties of chemical bonding. Nanoscience is not mechanics but chemistry. Drexler did not know what he was talking about.

Drexler had run into this kind of doubt early on in his work at MIT, where one of the faculty sneered that his ideas showed "utter contempt for chemistry." Drexler thought the faculty member was wrong and pushed ahead with his vision of physically manipulating atoms. Smalley now stepped in where Drexler's MIT professor had left off and challenged the idea of molecular manufacturing on the basis of some rather un-nano-sounding problems. He called these the problems of "sticky fingers" and "fat fingers."

The "sticky fingers" problem was the worry that the atoms and molecules a nanotechnician was trying to move would stick to whatever mechanical devices were being utilized to position them. This would happen because, at these scales, atoms that are not bonded to others are attracted to each other by so-called Van der Waals forces. It would be hard to position atoms precisely because the tools being used to position them would struggle to let them go. Feynman had anticipated this very problem in his 1959 lecture, proposing, "It would be like those old movies of a man with his hands full of molasses."

The "fat fingers" problem suggested that after you get into the business of actually moving atoms around with mechanical devices, it turns out that, contrary to Feynman's lecture title, there is in fact *not enough* room at the bottom to control the number of atoms that would be flying around in any given chemical reaction. Reactions do not involve single atoms but clusters of them, demanding dozens of "fingers" to control even the simplest of manipulations. Smalley challenged Drexler—and implicitly the claim made by Feynman—when he suggested that there simply is not enough room at the nanoscale for this many fat fingers.

As far as Smalley was concerned, the vision of physically controlling atoms was the wrong one. He chided Drexler by declaring that chemistry was more like love. It required a certain complex "dance" involving motion in multiple dimensions and the right blend of attractive forces and chemical connections. It was not something that could be done forcefully according to plans imposed by mechanical devices from the outside. "Fingers just can't do chemistry," Smalley said.

Drexler replied testily to Smalley that although sticky fingers might be a problem for atoms and researchers eating donuts, it was not a problem when moving molecules around. Biology proved this to us every day. Biology, in fact, was the proof of the whole concept that had first inspired Drexler to follow the path of molecular manufacturing.

The debate between the two heated up in a series of published articles and open letters. Smalley bluntly accused Drexler of failing to understand chemistry. Drexler responded that Smalley obviously did not grasp biology. Smalley pointed out that if biology was the proof that this could happen, then water would be a required medium for all future nano manufacturing. Molecular manufacturing, in other words, could only be "wet." Drexler told Smalley that he had "an inadequate grasp of the proposal" and was "confusing the public." Sensing that the dispute had turned personal, Smalley replied that Drexler was "scaring our children," an allusion to Drexler's global ecophagy blunder. The increasingly vitriolic dispute drained away even more of the scientific optimism surrounding molecular nanotechnology.[4]

In the years immediately following this public spat, many observers in the nanotech community found themselves frustrated by the tone and the egos they saw on display. They found the debate to be a distraction that did not have a lot to do with actual nanotechnology research. Most of the applications of nanotechnology thus far had depended on the innovative use of manufactured nanomaterials and not on the construction of futuristic nanobots. The highly speculative visions of molecular manufacturing seemed to many nanoscientists to be a fruitless sideshow. The idea of molecular-sized machines conducting precise surgical tasks or manufacturing commissions existed only in Drexler's dreams. It remained about as silly as the idea of nanocars running around delivering very small pizzas. Why waste precious scientific time, resources, and credibility debating such matters?

Although this frustration is understandable from the point of view of serious scientists wanting to get on with their nanotech research, to dismiss the Drexler-Smalley debate as pointless would be to pass over something that has huge significance for someone wanting to understand the sort of intervention into the natural world nanotechnology signifies. Despite its shortcomings, the Drexler-Smalley debate highlighted one of the most important implications of nanotechnology as a practice of the Synthetic Age. Smalley thought Drexler did not understand chemistry, and Drexler thought Smalley did not understand biology. In nature, both biology and chemistry make their fundamental moves at the nanoscale. Whoever might be judged to have won or lost the debate, the battle itself reveals something important about what nanotechnologists are really doing. Engineering materials and devices at the atomic and molecular level means that *Homo faber* is attempting to deliberately adjust long-established blueprints handed down through physics, biology, and chemistry in the hope that this will allow us to put the surrounding world to more effective use.

Although nanotechnicians would not claim to have gained the ability to rewrite the laws of nature, they clearly are learning how to work at the margins of those laws in ways that open up some startling new horizons. If the advocates of nanotechnology are to be believed, biochemical processes can be coopted and pushed in directions they have never been pushed throughout the earth's history. A whole new array of possibilities opens up. Call these possibilities Biology 2.0 or NextChemistry or perhaps Synthetic Physics. In each case, researchers are exploring portions of earthen territory to which our species had no previous access.

As Drexler's parable about grey goo makes clear, there are reasons to be cautious about accessing this territory and meddling with the long-established patterns found there. Efforts to completely control the world we create have a propensity to fall slightly short. Things happen that we did not anticipate. At the macro scale, materials get fatigued, unpredicted chemical reactions occur, and bizarre chains of events get linked together. Social uncertainties also have a habit of coming into play.

Former US Secretary of Defense Donald Rumsfeld famously warned about the hidden dangers of the "unknown unknowns." Nanotechnologists might do well to remember how these unknown unknowns can linger, however exhaustively scientists attempt to root them out. This is likely to be especially true in the unfamiliar realm of nano.

A philosopher of technology from the U.S. rust belt named Steven Vogel has captured this feature of the world by pointing out that constructing anything in the physical world involves accepting some loss of predictability. As soon as you make an idea real by building a device or structure, you have immediately relinquished a tiny portion of control over what that product will do.

This is a basic truth about artifacts. The physical world is filled with elements of unpredictability that get baked into the things we build. Even the best-constructed artifact retains just a little bit of wildness that forever has the potential to come back and haunt us. The corroding rebar in a bridge, the sudden crack in the airplane's hydraulic system, the undiscovered bug in the computer network: matter is never 100 percent stable. This is already true of artifacts that are relatively well-contained. But if artifacts are built to roam free and self-replicate, then this inherent wildness becomes a rapidly escalating concern. Grey goo provides a stark reminder of this truth.

Vogel's warning echoes similar concerns raised by others who have thought about the impacts of particular technologies. For example, a British professor named Keekok Lee wrote a book titled *The Natural and the Artefactual*. Although the work did not receive much attention outside of narrow philosophical circles at the time of its publication, it made an important point worth bearing in mind as we move into an increasingly Synthetic Age.

Lee expressed caution about "deep technologies" like nanotechnology that reach down into the very nature of things to reconfigure them for human purposes. Part of the problem for Lee was that this could be risky for human health and well-being. Our bodies are simply not used to such materials. Nor is the environment. Anticipating Vogel, Lee also suspected that we would fail to predict everything about the behavior of the products of deep technologies. Nanotechnology and technologies like biotechnology are, for Lee, inherently risky because of how far down into the structure of the world they take us.

There is, however, an additional concern about the danger of a big shift Lee detects in values and meaning. She sees nanotechnology as "nature-replacing" in the sense that it takes the stuff that nature provided and trades it out for something that humans have determined better serves their demands. Such actions, Lee thinks, have a moral dimension to them. Nanotechnology replaces something we have come to rely on—and perhaps, grudgingly, to respect—with something entirely artificial. It is a manipulation of something fundamental that impoverishes both ourselves and the world. We have transferred an important realm of being from the category of the natural over into the category of the artificial.

Lee's worry about nature being "replaced" by nanotechnology may be a slightly exaggerated fear if she means it literally. Short of a rapidly spreading grey goo, there will always be a healthy supply of natural, biological nature to be found. The world will retain a complement of ferns and waterfalls, beetles and sparrows, mountain lions and octopuses, all of which will continue to make the place more interesting and lively for us, whatever devices nanotechnologists dream up.

Yet Lee's concern about there being a moral dimension to deep technologies captures something philosophically important. If *Homo faber* really can reconfigure matter at such a fundamental level, then we can start to build a world that increasingly departs from the world of the past. Using techniques developed at the nanoscale, the boundaries provided by the very *nature* of nature will limit us less. Matter will increasingly be reconfigured from the atom upward so that the material world can serve us better. This does not simply mean that the world will be increasingly full of the endless different artifacts that we tend to fabricate as our wants relentlessly expand, although that will surely happen too. It means that the world will be full of an increasing number of *kinds of* things and types of matter that are thoroughly human, rather than natural, in their origin.

To many, this sounds like *Homo faber*'s dream. No longer limited to fabricating things out of the clay that the earth provides, humans could design their own clay in order to create a world that will do their bidding even more effectively. Machines and material structures that were previously considered to be physically impossible may become part of daily life. Humans could learn to think radically out of nature's box.

But in the process of doing this, something basic about our sense of the enveloping world and the limits it places on us starts to change. Artifacts used to be understood as things built out of a finite range of existing materials pulled from nature—materials such as woods, ores, liquid hydrocarbons, precious metals, and various familiar chemical elements. These materials could take us only so far before they posted a sign that said "No further!"

In the era of nanotechnology, these limits change. Manufacturing at the atomic and molecular levels means rethinking the whole idea of nature as a fundamental and limiting substrate on which humans impose their designs. Artifacts become things intentionally built out of matter *that has already been intentionally built.* Artifacts are thus twice artifactual, in both the final product and in the material out of which these products are constructed. We would be living in a world *deeply* remade, a world whose ability to limit us is increasingly being removed. This explosion of possibilities is both intensely exciting and slightly disorienting at the same time.

Pushing past these limits certainly offers something new and perhaps it is something worth pursuing. As Drexler promised, we could be looking at a future of "radical abundance." This is a future worth cautiously exploring. But at the same time, we should be intensely aware that it also marks a step into the unknown.

Nanotechnology could mark the point at which the best of science fiction finally becomes reality. It might signal the transcending of material limits, the expansion of creative possibilities, and the unlocking of numerous new economic opportunities. Alternatively, the development of nanotechnology could mark the moment when humanity starts rearranging our surroundings to such an extent that the familiar world in which our species has evolved becomes entirely alien to us. Within nanotechnology, there are reasons for both hope and for fear.

The dreams first articulated nearly sixty years ago by Richard Feynman represent merely one of the many entry points into the new world we are creating. Nanotechnology signifies the first of a number of ways of reengineering the world that characterize the Synthetic Age. Although nanotechnology's main focus is the abiotic parts of nature, other technologies focus on the natural world's biological elements. When one has gained the ability to work at the nanoscale, one has automatically gained the ability to operate at the scale of DNA. It is no surprise, then, to find a different breed of futurist trying to find interesting ways to tinker with the genetic elements that drive the living world.

3 DNA on Demand

Around two decades ago, just six months into the new millennium, the world celebrated the news that the human genome had been mapped. A joint press conference by U.S. President Bill Clinton and British Prime Minister Tony Blair announced the success of a public-private partnership in creating a draft sequence of all the rungs of the beautiful double-helix ladder of human DNA.

The Human Genome Project was completed ahead of schedule and under budget, thanks in part to the contributions of a private firm that joined the government-sponsored project in its later years. Free-marketers celebrated the fact that this private company, Celera Genomics, had brought to the table a more efficient technique for reading the all-important sequence. Comparing the genome map to other great territorial maps of ages past, President Clinton declared, with one of his trademark grins, "Without a doubt, this is the most important, most wondrous map ever produced by humankind." The president stated that, with a discovery this profound, "we are learning the language in which God created life."

It was certainly an auspicious achievement, one that demonstrated enormous patience and technical prowess. The human genome contains about 24,000 genes. These genes are made up of slightly more than three billion pairs of the nucleobases (adenine, cytosine, guanine, and thymine) that form the rungs of DNA's famous ladder. To map the genome, each base pair had to be identified and located in the right position on the long and spiraling genome.

Even before any reading of the nucleobases took place, small fragments of DNA had to be transferred to bacterial cells that acted like Xerox machines so that multiple copies of each fragment could be made. This copying technique ensured that scientists could be confident about the order of the letters they were reading. Under the best of conditions, gene sequences can be interpreted only in short stretches, so multiple overlapping sequences had to be mapped, and comparisons between the different stretches made. As each fragment was confirmed, a catalog could start to be pieced together for the entire genome. The whole three billion base pair sequence had to be checked and rechecked multiple times.

After nearly a decade of intense labor, the work of thousands of genomicists from more than eighteen countries was complete and ready to be celebrated. A reliable draft of the genome was made available, and smiling politicians could strut around giving everyone pats on the back and bask in reflected glory.[1]

The achievement was indeed historic. We know this because it was not only politicians who turned hyperbolic in their commentaries. Even scientists waxed uncharacteristically lyrical about how this was much more than simply a new set of facts to be reprinted in a biology textbook. Francis Collins, the director of the project, saw the decoding as providing an invaluable lens on the world. Looking backward, Collins declared, the genome told "a narrative of the journey of our species through time." Looking forward, the new knowledge promised "a transformative textbook of medicine, with insights that will give health care providers immense new powers to treat, prevent and cure disease." Prime Minister Blair, tapping into the enthusiasm, stated that the discovery marked not just the beginning of a new generation of medicine but the crossing of a "frontier" and a "new era" of human existence. Human life had been distilled down to its biochemical essence. The genetic makeup of our species had become a readable text ready for a whole new level of investigation and analysis.

Although there is no doubt that decoding the human genome has provided the potential for advances in all kinds of medical and diagnostic procedures, the years since the completion of the project have revealed numerous complexities that have dampened some of the initial enthusiasm. Mapping particular genes to particular diseases and behaviors is not as easy as trying to match pairs of turned-over playing cards. A whole host of complexity and happenstance surrounds the role played by genes in determining the creatures we become.

For one thing, in addition to the DNA found within the cell's nucleus that the genome project mapped, there is also DNA in the cytoplasm outside the cell's nucleus that has a significant influence on how humans develop. This latter variety, known as mitochondrial DNA, was never part of the purview of the Human Genome Project.

Nor is it all about genes. It has long been appreciated that a person's future is influenced by a combination of both genes and environment (nature and nurture). Genes can do only so much on their own. The environment in which a person is raised and lives their life exerts a strong influence on how, whether, and when those genes are turned on and off.

Nurture, it has recently been discovered, is also relevant not only to the person currently being nurtured. Although many factors cause genes to become active at different times in a person's life, it appears that events today also can turn genes on and off in later generations. Studies conducted on isolated but well-documented populations in Sweden have revealed that stresses endured by parents, such as deficient diets caused by periodic failures of harvests, can change the expression of DNA in future generations. DNA, in other words, appears capable of being "traumatized" by its environment with the effects not being felt until an intervening generation or two has passed. U.S. scientists have found something similar in the descendants of Holocaust survivors. This "genetically inherited trauma" means a grandchild can have a greater susceptibility to ailments like diabetes and heart disease even if nothing particularly unusual showed up in the earlier generation that actually experienced the stress. In a distant echo of Lamarckian thinking, individuals appear to be able to pass on to future generations, through their genome, the consequences of something that was experienced during their lifetime.

Another wildcard is the influence on health and disease of the hundred trillion or so single-celled microbes that inhabit our bodies. From our mouths to the insides of our colons to our toe nails, vast numbers of these simple organisms hitch rides on our bodies throughout our lives, mostly keeping us healthy but occasionally bringing us down with a bump. Genetically, we are more microbe than human. The total amount of genetic material contained in these microbes is up to a hundred times the amount contained in our own cells. Microbes affect our smell, our mood, and our behavior, influencing who we hang out with and probably whom we mate with. It is impossible to become our fully human selves without the right mix of microbes accompanying us on each stage of the journey. Because of the enormous influence of this *microbiome*, the object of natural selection for our species is now thought to be less the human genome itself than the whole human ecosystem. Medicine may be as much ecology as it is genetics.

A whole field, epigenetics, has been developed to look at how cells might read genes differently depending on factors external to the genome itself. As Diane Ackerman has put it, "Epigenetics is the second pair of pants in the genetic suit." And it is clearly a very big pair of pants. Although the human genome contains only 24,000 genes, the epigenome includes millions of factors that influence human development. The genome itself, in other words, holds only a few of the cards. Sequencing the genome, in the end, stood no chance of telling the complete story. When philosophically minded observers despaired that the Human Genome Project threatened to remove the mystery and poetry of life by reducing humans to a chemical blueprint, they far underestimated the complexity of what goes into making us who we are.

Although the Human Genome Project did not unlock all the mysteries of the human body in one fell swoop, it did contribute mightily to an additional line of transformative research. This line of research will, in the end, probably do more to shape the Synthetic Age than any mapping of an existing genome could ever do. Rapidly improving gene-reading technologies that were refined during the Human Genome Project started to push open a different but significant door on which certain ambitious commercial interests had been persistently knocking.

Celera, the private company that worked alongside the British and American governments on decoding the human genome, was never particularly focused on the connections between human genetics and human health and behavior. Those connections just did not interest them much. What everyone else thought was the central motivation to study genomics was, for this company, a bit of a sideshow. Celera's highly ambitious founder, who sat alongside President Clinton at the White House celebration, had more radical goals in mind.

• • •

J. Craig Venter was born in Salt Lake City to a hard-drinking, hard-smoking Mormon father. In his son's early years, Venter senior suffered the indignity of being excommunicated from the Mormon church. In order to escape from this public shame, the family moved from Utah to a working-class area outside San Francisco, where the younger Venter reported that he felt liberated by the expansive possibilities provided by the coast. Happy to have left the desiccated basins of the mountain west, Craig quickly discovered in California a lifelong love for the ocean.

School, by contrast, was not one of his passions. Even though he found himself fascinated by his shop class, Venter was never a particularly good student, and he limped through high school with a series of unimpressive grades. He much preferred to spend his time on the ocean, lying on his surfboard or swimming for hours along the beach. Nobody at high school thought Venter would amount to much.

As a young adult, Venter was drafted into the Vietnam War and served in the Naval Medical Corps. In southeast Asia, he worked at a field hospital where he helped treat critically injured servicemen. During his tour, he was exposed to the horrific aftermath of the Tet Offensive. His dark experiences in Vietnam were pivotal to Venter's future. At a particularly low point during his deployment, he came close to committing suicide by swimming as far as he could into the ocean with the intention of not returning. Circled by a shark a mile out from shore, he had second thoughts and swam slowly back. He became determined to survive the war and return home. He devoted himself to the idea of fixing the inadequacies in the field treatment of injured soldiers he had witnessed in Vietnam by studying medicine.

Returning to the United States and entering university, Venter quickly found himself more interested in biochemistry and physiology than in medicine. His childhood interest in shop class meant that at heart Venter was a mechanic and a tinkerer. At the same time, he also was developing a hunger for entrepreneurship.

In the study of genomics, Venter found the perfect outlet for his passions. After finishing his PhD, he took a university job in Buffalo before moving to the National Institutes for Health (NIH) for an eight-year stay during which he developed a new technique for identifying individual genes and their functions. In 1992, he left NIH and founded the Institute for Genomic Research (TIGR), a nonprofit private research organization where he continued working on reading and interpreting genomes, something he now realized he was really good at. While at TIGR, Venter helped refine something known as the *shotgun* technique for sequencing DNA. By repeatedly shattering genomes into numerous shorter lengths and identifying the fragments, Venter found that computers could be used to match up the thousands of reads and map the extended sequences. By deploying huge computational resources, TIGR soon became the most efficient reader of genomes in the world.

Somewhat bemused by the slow pace of public efforts to sequence the human genome, in 1998 Venter became president and chief scientific officer of a new gene-mapping company, Celera Genomics, in part to help bring the shotgun technique to the human genome. He figured that it should be possible to map a human genome in three years using this method, compared with the ten years that the public project had planned. Celera at one point was looking to profit from this work, although it recognized the growing sense among scientists and the public that the sequence of the human genome was a public good and should not be cornered by any one company for money making.

Celera succeeded in its sequencing goals, and in order not to eclipse the public project, Venter shared the podium with project leader, Francis Collins, at the White House ceremony in 2000. Knowing that his work with the human genome was done, Venter pivoted. Leaving Celera abruptly, he returned to his longtime scientific collaborators at TIGR. There the world's best genome sequencer quickly refocused his attention on what had always seemed to him a much more significant target.

When genome sequencing got off the ground in the late 1990s, TIGR was the first to sequence the entire genome of a completely free-living organism, the bacterium *Hemophilus influenza*. Shortly after sequencing this genome, the group successfully decoded one of the smallest genomes known to exist, *Mycoplasma genitalium*, a bacterium that lives in the human urinary tract and helps spread sexually transmitted diseases. After sequencing *M. genitalium*, the scientists at TIGR carefully washed their hands and moved on to other minuscule genomes, sequencing more than fifty of them in the next few years.

Venter called this sequencing of small genomes his "Minimal Genome Project," and initially this focus on tiny organisms seemed to present a puzzle. Why would a private company spend good money and time focusing on such simple organisms when there are much more complex and potentially rewarding organisms out there? As techniques improved, most research groups moved on from bacteria to sequence progressively more sophisticated organisms such as frogs, mice, and chimps. Mapping genomes similar to humans' for their medical implications was where everyone anticipated all the big money would be made. Venter's team at TIGR, however, was not interested.

The explanation for this comes straight out of a Plastocene playbook. Venter's goal was not only to read genomes but to build them.

• • •

The field of synthetic biology was just emerging when the Human Genome Project began. Synthetic biology rests on the idea that biology should become more like engineering. This means learning how to design, build, manipulate, and replicate biological devices with precision and certainty. A genome is ultimately no more than a particularly interesting chemical structure with a certain arrangement of phosphorous, carbon, oxygen, hydrogen, and nitrogen atoms. If a researcher can understand what the chemical organization of a genome is and what the different pieces do, she should be able to take it apart, put it together again, and juggle some of the interesting sections. With enough patience and ambition, she ought to be able to engineer new genetic combinations. Synthetic biology is about expanding and deepening techniques so that genomes can be built to order. It promises a DIY version of biology in which humans, not evolution, call the shots.

The field of agricultural biotechnology showed the value of moving individual genes from one organism to another to create desirable traits. What if you could go beyond moving the odd gene or two into different species and start to swap out or construct extended gene sequences? Perhaps you could create not just traits but whole biological systems that could produce valuable things for you.

For example, what if you could identify all the genes responsible for producing a certain chemical in one organism and relocate them to a more user-friendly second organism, creating what would amount to a biological factory? If you needed to transform this chemical into something else, maybe you could add the appropriate genes from a third organism to manufacture a new type of biological production unit. Synthetic biologists could start building complicated but useful gene sequences the likes of which nature had never seen before.

Venter's thoughts now drifted even more ambitiously to the manufacture of living organisms themselves. Because genes are made up of relatively simple chemicals, there is no reason, thought Venter, that nature should be the only one constructing viable DNA chains. Humans could do it, too. Instead of using synthetic biology to construct useful sections of genomes, it may become possible to synthesize the whole thing. One could use the techniques of synthetic biology to design and manufacture whole organisms in the lab instead of waiting for nature to produce them.

This audacious possibility quickly became Venter's primary goal. In 2006, he formed an umbrella organization called the J. Craig Venter Institute (JCVI) that brought under one roof the growing number of his research and commercial interests. Within JCVI, some of the original TIGR researchers, now joined by other leading genomicists attracted to what had become one of the world's leading scientific laboratories, set to the task of building a viable genome from scratch, using its constituent chemicals.

Venter knew that it would be a spectacular trick for humans to show they could manufacture a living organism's whole genome from its component chemicals. From a philosophical perspective, humans would be entering an entirely different realm of endeavor. The genome of a living organism was something that *Homo faber* had never built. Achieving this would essentially take Bill Clinton's promise about scientists learning God's language one step further. *Homo sapiens* would not just show itself to be capable of reading God's language. It would show that it could pick up a pen and write it.

But writing copies was not the only thing that Venter had in mind. If humans could successfully construct genomes, they would no longer need to be satisfied with copying the genome of whatever organism nature happened to have come up with. They could design new ones from the ground up that were more interesting, more useful, and perhaps—Venter's entrepreneurial mind was now buzzing—more profitable. Instead of building inanimate machines to perform useful functions, humans could build *organisms* for similar purposes. Organisms never before seen in nature could be created entirely to serve us. It would be a remarkable, if slightly Frankensteinian, achievement.

In this dream of Venter's, it is possible to detect a certain overlap with the dream of molecular manufacturing in nanotechnology. The overlap is real: the rungs of the DNA ladder are about two nanometers across, so by definition DNA synthesis would be an activity that takes place at the nanoscale. As we saw above, molecular manufacturing had already taken a biological turn. Although DNA bases would not be assembled into sequences using rods and ratchets as nanotech's molecular manufacturing promised, technicians would have to take the same degree of care to ensure that the nanosized bases were put together in the correct order. As Richard Smalley had predicted for nanotechnology, all this would have to take place in aqueous solution. In other words, the whole endeavor would be a form of *wet nanotechnology*.

What makes synthetic biology philosophically different from the proposed molecular manufacturing in nanotechnology is that the artifacts to be designed in synthetic biology would be not simply machines but actual living organisms. Successful biological organisms are remarkable structures. They fuel themselves, they repair their injuries, and they reproduce each subsequent generation without assistance. Evolutionary pressures have shaped organisms to perform precise functions with efficiency and reliability. If humans could design organisms to execute operations that overlapped with human interests, then they would quite possibly have at their disposal the most effective and low-maintenance machines imaginable. This promised to create an industrialist's nirvana.

Having watched the public's reaction to genetically modified organisms, Venter knew that synthetic biology was bound to raise concerns among certain segments of the public. It was, after all, an attempt to create life in a lab. Before a synthetic microbe could be put to any useful purpose, a number of precautions would have to be taken. Learning from Eric Drexler's mistakes, Venter recognized that care would have to be taken not to arouse fears about synthetic microbes going on a bacterial rampage. Just as with nanotechnology, it seemed plausible that these intentionally designed biobots could multiply uncontrollably, although in this case, the rampaging goo would be green and not grey. Venter would have to provide assurances that he could prevent their escape into the wider environment. He would have to conduct studies to show that the risks to human health from synthetic genomes were minimal. He also would have to take into account the biosecurity issues raised by the possibility that synthetic life might fall into the wrong hands.

From the start, Venter set out to make it clear that he took all these ethical issues seriously. He created a policy group at JCVI that teamed up with the Center for Strategic and International Studies and the Massachusetts Institute of Technology to investigate the ethical concerns generated by fears about synthetic organisms running wild. They looked at various dangerous scenarios and devised principles that should be followed to minimize the dangers. The ethics, Venter insisted, was under control.

• • •

Venter had always been drawn to cutting new paths, but the conceptual thresholds to be crossed by this work were nothing short of astonishing. His work was located in biology but at the point where the line between biology and philosophy starts to blur. In a similar way to how nanotechnology pushes into the depths of physics and chemistry, in synthetic biology humans would be coopting some basic mechanisms of biology for their own usage. The living world would no longer be a world that resulted from three and a half billion years of biological history prior to the arrival of humans. It would be a world we had shaped and designed ourselves to meet our needs.

Huge questions about meaning and value loomed. With the prospect of synthetic biology, the distinction between the living and the artificial starts to blur. Two normally separate categories—life and machine—would blend in a way that they had not blended before. The machines constructed by humans up to this point had all been abiotic. They had not self-replicated or looked after themselves, they generally needed an external power source, they tended not to have been made out of organic molecules, and they typically depended on operators to press a "start" button in order to initiate their action. Outside of Drexler's grey goo nightmare, there had been no danger of these machines ever having a life of their own.

This was all about to change. If Venter succeeded, some machines designed by human technicians would now be *living* machines able to survive and protect themselves. This would be a new frontier for the earth and its systems. A new relationship would emerge between humans and the living world. Our species would become creators of life forms as they started to manufacture a suite of what would essentially be *biotic artifacts*.

All of this sounds dramatic. It has the appearance of a significant rupture from all of past human history and the beginning of an unprecedented synthetic future. But is the idea of a biotic artifact really new? Some observers of the biological sciences have suggested that this is all too familiar territory. Domestic cattle and sheep, they say, are already a type of "living machine" manufactured to suit human needs. Those creatures have been designed through careful breeding to perform a function that humans find useful, such as producing milk, growing wool, or providing a side of beef. The same is true for big-eared wheat and corn plants. Many of these doctored organisms are self-replicating and, to some degree or other, self-maintaining. With domesticated plants and animals, humans already appear to have shaped the living world to meet their needs and to make them money. A trip to the farm is all that is needed to encounter a range of existing biological machines.

There is certainly some truth to this claim. Domesticated cattle are undoubtedly living organisms shaped by intentional human design. But there remains a significant conceptual difference between a synthetic bacterium and a carefully bred farm animal. A cow or a sheep is not a machine in quite the same way as a designer microbe would be. A domestic animal is made out of some fairly natural raw material—namely, its parents. Sheep remain closely related to their wild ancestors and have a long biological history that a microbe designed from scratch would completely lack. A cow is carefully bred so that an already useful organism gains additional value. By domesticating animals, humans carefully build on some of nature's own discoveries, adding a slight adjustment here and a gentle shaping there so that the existing species better meets our needs.

With a synthetic microbe, by contrast, the organism is created from scratch expressly to serve the human interest—not just by cross-breeding a very woolly specimen, for example, with a very muscled one but by building the exact genome that would best serve human purposes. This type of manipulation would not be merely a *shaping*; it would be a *creating*. The synthesized organism would be an artifact through and through, built to a specific design. The creation would take place, moreover, in a sterilized lab at the hands of scientists in white coats using technologically sophisticated tools rather than in a damp field filled with bleating farm animals and the stench of manure.

Synthetic biology and the idea of constructing genomes from scratch looks like a clear example of what Keekok Lee calls a "deep technology." It reaches far into the workings of nature and promises such radical changes to the concept of "life" that it differs substantially from anything that has come before. Like nanotechnology, synthetic biology is an engineering tool fit for a Synthetic Age. But instead of simply tweaking aspects of nature's physical and chemical structure, synthetic biology tweaks life itself. Even more so than nanotechnology, synthetic biology crosses an important line. It turns humans into a new and more powerful type of creator. We would be designing and building the living world around us, surrounding ourselves with the monsters of our own making.

And unlike molecular manufacturing in nanotechnology, as we are about to see, synthetic biology has already made significant progress toward its goals.

4 Artificial Organisms

The advances to date in synthetic biology have occurred through a series of incremental steps. Before attempting to create an entire organism from scratch, synthetic biologists worked on developing strings of useful gene sequences. A stepping stone to a wholly synthetic microbe is a microbe that contains extensive sections of DNA that are built to spec in the lab. If you are doing it right, these sections then can be inserted into a host organism to perform a valuable function. The most notable example of this so far has been the engineering of a biological system that can produce an essential precursor of the antimalarial drug artemisinin.

In the early 2000s, a team in California led by a biologist named Jay Keasling deliberately altered the DNA of a yeast cell through the introduction of a significant amount of genetic material from the wormwood plant. Wormwood, long used in traditional herbal medicine, was scrutinized for its antimalarial properties by Chinese scientists in the early 1970s. The discovery of effective clinical methods to extract artemisinin from the wormwood plant came as a direct result of an order from Mao Zedong to figure out how to stem the devastating impacts of malaria on Vietnamese soldiers during the war with the United States. Although the drug extracted from the wormwood plant was effective, it remained both slow and expensive to produce. For many years, there was an ongoing effort to figure out how to manufacture the drug synthetically in a lab.

By importing the set of genes responsible for performing the antimalarial chemistry in wormwood, Keasling and his colleagues managed to engineer yeast cells to produce artemisinin much more efficiently than wormwood plants can do it. It is a dramatic genetic trick. After the genetic import, at least one portion of the yeast cell is no longer really doing what yeast cells do. It is doing what wormwood cells do.

The type of manipulation that Keasling's lab achieved far exceeded anything that had occurred in traditional genetic modification (GM), for example, in the creation of Bt cotton plants or Roundup Ready soy plants. To make the useful antimalaria chemical, several genes in the yeast had to be turned on and off to allow this novel function to take place within its walls. The inserted wormwood genes also had to be tweaked so that they could perform effectively in a different host organism. When Keasling's team figured out the engineering, the yeast cell became a living production facility for the antimalarial agent, something the yeast cell would never have imagined when it contemplated a life ahead spent fermenting beer and leavening bread. This type of metabolic engineering essentially sites a biological factory inside the body of another organism. Using these methods, other useful medicines (such as synthetic antibiotics and synthetic vaccines) are also within the realm of possibility.

The goals of metabolic engineering projects like Keasling's are being advanced by the creation of an official inventory of useful genetic parts. These parts have become known as *biobricks*. Each biological brick is known to perform a certain useful function. An international biobrick registry administered by the Massachusetts Institute of Technology is available to any researcher in the world. This registry contains over three thousand useful gene sequences in standardized formats, any of which can be ordered online (as if through Amazon). The biobrick registry is essentially synthetic biology's digital warehouse, which manufacturers can call up when they need something for the biological machine they are building. Unlike most industrial warehouses, the biobrick registry is not for profit. It also is completely open source. It has been created specifically to help advance an emerging industry.

Metabolic engineering, impressive as it is, is still just a halfway house. The real goal for people like Venter remained the entirely synthetic genome. As Keasling perfected the process for engineering his antimalarial drug, researchers on Venter's team were closing in on their target of building a wholly synthetic genome.

Only three years after the completion of the Human Genome Project, a Venter research group made up of Clyde Hutchinson, Cynthia Pfannkoch, and Hamilton Smith took the important first step of moving from simply reading genomes to building them. They did this by synthesizing the genome of a virus known as PhiX174 from laboratory chemicals. Although a notable milestone, viruses are not thought of as free-living organisms because they need a host organism in order to survive. More had to be done.

In 2007, four years after the success with the virus, Venter's team figured out how to replace the genome in one bacterium with the genome of another and have the introduced genome take over the operation of the cell. This bacterial genome transfer was another important precursor to what lay ahead. Learning more about genome synthesis and translocation as each month went by, in 2008 the team synthesized the entire genome of *Mycoplasma genitalium*, the bacteria of the urinary tract it had successfully decoded back in the 1990s.

Even though *Mycoplasma genitalium* had a remarkably short genome for a living being, it still contained an awful lot of chemical structure. Numerous technical obstacles had to be overcome to create such a long strand of DNA. In addition to the brittleness of the increasingly lengthy fragments of DNA, there was the sheer number of nucleotides—582,970 pairs—that had to be correctly lined up. The stitching together required the assistance of those friendly yeast cells that Keasling's team had already been using for the antimalarial drug. Yeast turns out to be remarkably hospitable to bacterial DNA.

The announcement of the fabrication of the *Mycoplasma genitalium* genome took synthetic biology across a new threshold. This was the first time the whole genome of an independent organism had been built from its constituent chemicals in a lab. Unlike viruses, bacteria can make and store energy. They also can replicate independently of any host. This meant that scientists could now replicate the genome of a free-living lifeform in the lab and could do it entirely independently of any natural processes. The gamble Venter took in sticking with simpler organisms after the completion of the Human Genome Project was starting to look like it might pay off.

Notable as this achievement was, a string of DNA is not yet an organism. To create a synthetic organism, the genome Venter's team had synthesized would have to be placed into a friendly host so that the instructions it contained could be put to work running an actual organism. The 2007 successful translocation of a bacterial genome had showed that it was possible to insert a nonsynthetic genome into a different bacterial cell and have the new material take over. To create the first truly synthetic cell, they had to perform the same trick with an entirely synthesized genome.

The team found itself forced to switch from the *Mycoplasma genitalium* genome to a larger bacterium, *Mycoplasma mycoides*, because of the advantages offered by its faster replication. Having successfully synthesized this longer genome, what remained was the challenge of transplanting the synthesized material into the bacterial host and then "booting up" the host cell so that it ran off the inserted DNA. The chosen host was going to be yet another type of bacterium, *Mycoplasma capricolum*.

The technical challenges meant the process was pretty slow going. The cell envelopes of bacteria generally are not highly fortified, and so without precautions in place, bacteria would be trading DNA like cards at a poker game. This genetic promiscuity requires that, to protect themselves against the uptake of undesirable genes, numerous restriction systems are built into bacteria to fight off foreign DNA. Before the Venter scientists could insert the synthetic *M. mycoides* genome into its *capricolum* host, they had to find ways to circumvent each of these defenses. They also had to work hard at keeping the long and brittle genome intact in the process of transplanting it.

After a decade of work and about $40 million spent, the first successful synthetic genome transfer was announced in an article published in May 2010 by the Venter group in the journal *Science*. An entirely synthetic genome modeled on *Mycoplasma mycoides* had been inserted into a *Mycoplasma capricolum* host with the *M. mycoides* taking over the running of the cell. The new organism was named by the team *Mycoplasma mycoides JCVI-syn1.0* and was described proudly by Venter as "the world's first synthetic cell." It almost immediately began reproducing.

In order that its progeny could easily be distinguished from naturally occurring *M. mycoides* bacteria, the researchers encoded several genetic markers into a nonactive part of its genome. They included, with some geekish flair, a web address for the new organism and the James Joyce quote "To live, to err, to fall, to triumph, to recreate life out of life." It also included a quote from the nanotechnology pioneer Richard Feynman: "What I cannot create, I cannot understand." The media promptly dubbed the new organism *Synthia*.

The giddy reaction of the press and much of the scientific community made it clear just what a big deal this was. Venter himself claimed that booting up the synthesized genome of *M. mycoides JCVI-syn1.0* was as much a conceptual breakthrough as a technical one, calling it a "giant philosophical leap in terms of how we view life." He brazenly described the possibilities ahead as a "new phase in evolution" where one species could sit down in front of a computer and design another. Other enthusiastic supporters referred to the potential of synthetic biology as "Life, Version 2.0" and as a matter of "out-designing evolution." To many, this was a dramatic new responsibility for humanity. One practitioner, not shy of the divine role it implied, called it "the Regenesis."

Not everyone was quite as bullish. Some grudgingly pointed out that JCVI's cell was only semisynthetic because the constructed DNA had been inserted into a nonsynthetic bacterial host. Others found Venter to be too bombastic about the whole thing. In an echo of the testy Drexler-Smalley debate in nanotechnology, Venter's remarks prompted Jay Keasling to suggest (in answer to a question about the regulation of synthetic biology) that the only thing that really needed regulating in this new field was "my colleague's mouth."[1]

Despite the enormity of the success in creating the synthetic organism, Venter's Minimal Genome Project still was not done. He wanted to refine the techniques used to build *M. mycoides JCVI-syn1.0* in order to create the smallest possible genome that could function to keep a bacterial cell alive. Because evolution often takes a long and winding road to end up with a particular organism, every genome has genes that have become redundant and are inessential to life. Venter suspected his team could go smaller than *M. mycoides JCVI-syn1.0.* The Venter Institute filed patents for this future minimal synthetic bacterium, a life form it called in anticipation *Mycoplasma laboratorium.* Researchers then set to work to establish its genome by methodically removing what they determined to be the inessential genes in *Synthia.*

The idea of pursuing this smallest possible microbial genome was driven by its enormous potential for commerce. Possessing a viable minimal genome aligned neatly with the underlying goals of the biobrick registry. Such a minimal organism could be used as a living framework into which functional biobricks could be inserted. This most basic of organisms would then be like a biological factory floor on which a desired bioindustrial machinery could be sited. It was here, finally, that Venter anticipated big money could be made.

In March 2016, Venter researchers published a journal article showing that they had identified and then synthesized all of the 473 genes necessary for what they had determined to be this simplest living being. After successfully inserting this minimal genome into a bacterial host and booting it up, Venter claimed to have created "the first designer organism in history." Unlike *M. mycoides JCVI-syn1*.0, this was not just a copy of an existing genome. It was a tiny, entirely new form of life. In the year 2000, Sun Microsystems founder Bill Joy had predicted that "the replicating and evolving processes that have been confined to the natural world are about to become realms of human endeavor."[2] Through the process of designing and building a minimal genome, Venter's team had finally fulfilled Joy's prediction. The artificial form of life they had created did not emerge from evolution. It emerged from workings of the synapses of the human brain. Intelligent design typically refers to divine explanations of an organism's origins by Christians who are suspicious of evolution. In this case, for the first time humans had themselves become the intelligent designers of life.

Nearly twenty years after decoding what was thought to be the world's smallest naturally occurring genome in *Mycoplasma genitalium* and nearly half a century after returning from Vietnam, Venter had succeeded in building to his own design what was probably one of the smallest forms of life the planet had seen in many millions of years.

• • •

Venter is fond of suggesting in his public lectures that the world's first trillionaire will be the person who designs and produces at scale the first economically desirable synthetic organism. The uses to which such organisms might be put are significant. Multinational business interests are lining up to partner with Synthetic Genomics, JCVI's commercial spinoff. These corporations include British Petroleum, the agricultural giants Monsanto and Archer Daniels Midland, the pharmaceutical company Novartis, and the U.S. Defense Department's research wing, Defense Advanced Research Projects Agency (DARPA). The fossil fuel company Exxon Mobil also committed up to $300 million to a partnership with Synthetic Genomics in order to develop synthetic algae that would produce biofuel.[3]

With the mention of these sorts of alliances, environmentalists typically start to get nervous. But as with nanotechnology, ecologically minded observers have to concede that these biological mini-machines could perform a number of extremely desirable tasks. Venter peppers his trillionaire talk with frequent mentions of the environmental benefits of synthetic organisms. In addition to synthetic fuel, microbes could be designed to consume carbon dioxide out of the atmosphere in order to help solve the global warming problem. They could be constructed to break down cellulose more effectively to kick start biofuel production. A different sort of synthetic microbe could be designed to remediate pollutants on contaminated sites.

Synthetic organisms would have a number of inherent advantages over nonbiological machines built for the same task. Microbes are made out of some of the most abundant elements on the planet and would not require any expensive sourcing of parts. They are powered by ambient resources, are self-maintaining and self-repairing, have the potential for endless self-replication, and do not pollute in any traditional sense of the term. They are entirely organic and do not need to be disposed of when their useful life is over, decomposing naturally into their constituent components. You can see why a green entrepreneur might think there was money to be made. If synthetic microbes can dial back global warming and provide abundant carbon-neutral fuel to boot, how can environmentalists object?

A familiar line of objection focuses on risk. From the point of view of safety, one might wonder if adopting the role of creator of life is altogether wise. Of all the ways one can interfere with nature, attempting to achieve in a couple of decades of genomic research what it took nature three and a half billion years of biological trial and error to accomplish seems like it might be one of the most ill-advised. The ecological risks posed by synthetic microbial organisms could yet turn out to be significant. After they are set free in the environment, it is unclear whether they would ever be able to be brought back in. The worry about uncontrolled nanobots graphically portrayed by Michael Crichton could reappear in the form of a global epidemic of synthetic bacteria. It is worth remembering that fundamental to DNA is the fact that it undergoes random mutation.

There is, however, a different kind of objection that makes the skin on the necks of people like Keekok Lee crawl. A hint of it is given when advocates of synthetic biology start talking about "outdesigning evolution" and "reinventing nature." President Clinton inadvertently raised it in his speech to celebrate the end of the Human Genome Project when he referred to learning the language of life. To echo in biology what Drexler's professor told him in chemistry, it is the utter contempt for Darwin that is so shocking.

• • •

Humans have been in the business of disregarding the forces of natural evolution since the first crops were domesticated in the Fertile Crescent over eleven thousand years ago. By the time Gregor Mendel finished experimenting with his pea plants in the 1850s, these manipulations had a sound scientific basis in the principles of heredity. Dogs and pigeons manipulated by Victorian breeders to satisfy a range of aesthetic ideals showed that humans would not hesitate to push the animal form in directions that suited their taste. Darwin himself appreciated and learned from these practices. In the last four decades, since the DNA of an *Escherichia coli* bacterium was first deliberately manipulated in the lab in the 1970s by the biologists Stanley Cohen and Herbert Boyer, humans have shaped genomes not only by controlling reproduction but also more directly by adding and deleting specific genes using "gene guns" and other technical means. Humans have a long history of tweaking genomes to serve their purposes alongside whatever tweaking nature itself was doing.

None of this long record of genome manipulation, however, comes anywhere close to the radical rupture from biological history made possible by synthetic biology. The advent of this technology destroys what had yet remained a broadly Darwinian monopoly on explanations for the origin of life.

Before synthetic organisms, it always had been possible to say something reassuringly Darwinian about every organism on earth. Every living thing—from a lately discovered antelope in the Vietnamese forest to a cotton plant with a gene inserted from the bacterium *Bacillus thuringiensis*—always inherited an overwhelming percentage of its DNA from previously living things. With the exception of some bacterial and mitochondrial DNA moving horizontally between organisms within a generation, genomes have always been handed down vertically by ancestors through reproductive means. These ancestors had their own ancestors, who all had physical connections to still more ancestors stretching far back into the reaches of evolutionary time. Before the advent of synthetic genomes, there always had been a concrete genetic link between parent and descendant. This is why it has always been true to claim that all of life is descended from a common ancestor. For three and a half billion years, Darwinian principles of selection had served as a deep time anchor for every organism.

Controversial as they are for many of the public, even genetically modified organisms do not replace Darwinian selection in quite the same way that synthetic biology does. GM crops are today grown on more than 175 million hectares worldwide and have revolutionized agriculture. Despite the fact that the famous Indian anti-GMO activist Vandana Shiva has suggested that GMO stands for "God move over" and not "genetically modified organism," GM technology is in fact far more grounded in Darwinian history than she acknowledges. All the organisms engineered since Cohen and Boyer's breakthroughs have retained their causal connection to evolutionary history. As the name implies, GMOs contain only *modifications* of existing genomes, changes typically affecting far less than a tenth of a percent of the total number of genes in the organism. The bulk of the genetic material in these modified organisms is born entirely of the earth's long history. This is true of both the 99.9 percent of the genome that has not been modified *and* of the less than 0.1 percent that has—because the foreign gene is itself a product of its own evolutionary history (albeit in a different organism).

Although crop breeding and agricultural biotechnology have allowed human wishes to be incorporated into an organism's DNA, resulting in valuable changes to that organism's behavior, they have not threatened the fundamental Darwinian fact that all the DNA in a GMO still has its origins in an evolutionary past. Carefully bred hypoallergenic lap dogs still have progenitors linked to *Canis lupus* and beyond. The worst of the so-called Frankenfoods at the center of anti-GMO campaigns have always remained causally connected to the history of life on the planet. Edited genomes still have their identity physically rooted in ancient ancestors.

Synthetic biology for the first time completely severs this causal chain. Synthesizing an entire genome from its constituent chemicals as happened with *Synthia* and *Mycoplasma laboratorium* crosses a new conceptual line in the sand. A synthetic organism literally has no ancestors. The genome inserted into the bacterial host has undergone no descent through modification. Nothing has been passed down. Nothing is inherited.

Clyde Hutchinson, one of the Venter Institute's researchers, highlights this difference in his reflections on their achievement: "To me the most remarkable thing about our synthetic cell is that its genome was designed in the computer and brought to life through chemical synthesis, without using any pieces of natural DNA."[4] The genome originates not in nature but in test tubes. The synthetic genome, it might be said, is entirely postnatural. Diane Ackerman highlights the character of the change. In these new types of synthetic organisms, as Ackerman puts it, "digital nature replaces biological nature."

Synthetic biology picks up where nanotechnology's fabricating instincts left off and pushes things further. As synthetic organisms start to populate the earth, the evolution of life as we have come to understand it through Darwin is progressively left behind. For the first time, as Venter points out, humans can look out on the living world and find an organism whose DNA is forged not by Darwinian evolution but by human intelligence. Descent through modification at last has a competitor. For the first time, humans will become creators of nonhuman life. Instead of destroying lifeforms, humans will start adding to the planet's complement of life by designing completely new ones. Some see this as a triumph. Others see only an extraordinary arrogance.

• • •

In a 2003 book about the rapid advances he saw occurring in genome manipulation, the environmental writer Bill McKibben sought to awaken his audience to the high stakes involved. Never before had humans attempted to remake the biological world this fundamentally. It was a complete departure for our species and promised uncertain and disturbing times ahead. This was particularly the case, thought McKibben, if we employed these technologies on ourselves. Drawing a line in the sand against some of the most aggressive genetic technologies is absolutely necessary, McKibben insisted, if we are to remain human. The title of his book on the future of genome manipulation proclaimed "Enough!"

The moral heart of McKibben's plea was a call for restraint. Humans have come a long way through their use of technology, but they need to be able to recognize that some areas are better left alone. He saw considerable dangers in embarking on a Synthetic Age that threatens to usurp natural evolution. His words were rooted in a hope that this might yet become "the epoch when people decide at least to go no farther down the path we've been following." The decision to stop that advance, McKibben believed, is fundamentally a decision about who we want to be. In McKibben's eyes, it is a decision to choose humility over arrogance. It is a choice to "remain God's creatures instead of making ourselves gods."[5]

According to the line of thinking developed by Paul Crutzen, the Dutch atmospheric scientist who suggested that we decide what nature is and what it will be, McKibben does not understand the role now demanded of us. Genetic manipulation of the type that leads to synthetic organisms is exactly the right technology for a Synthetic Age. At this point, we have little choice but to start consciously engineering both the physical and biological worlds in the light of our species' irrevocable impacts on the planet. Crutzen has suggested there will be a special role for scientists and engineers to "guide society towards environmentally sustainable management" during the Plastocene epoch. Synthetic biology is just one of a number of technologies that will be required if humans are to create a planet that can adequately support us. Intervening dramatically into natural processes is a task that Crutzen suggests is simultaneously both "daunting" and "exciting."[6]

The debate between people like Crutzen and McKibben represents the difference between those who see the Plastocene as an opportunity to gun the throttle and aggressively take control of our surroundings and those who see it as a call to slow down and begin to rethink our level of interference with nature. Despite Crutzen's enthusiasm, there is no inevitability to building genomes. This *is* a big moment of choice about where we want the Plastocene to go. We can pause and ask ourselves what we are doing while weighing the risks and dangers of the path we appear to be on. We might marvel at the idea of using metabolic engineering to produce new and valuable medicines in laboratory settings. But we also might hesitate at the thought of releasing synthetic organisms into the environment, concerned about their potential to mutate and start behaving in ways that nobody had anticipated. That tiny bit of wildness that the philosopher Steve Vogel suggested lurks in everything we build should remain at the forefront of our minds when we consider sending synthetic organisms into the surrounding world to perform tasks for us.

As we have done with practices like human cloning, we could draw lines in the sand at particular thresholds. We could recognize boundaries that ought not to be crossed, either because they create too much risk or because they change too much about the world around us and, in the process, change too much about ourselves. Alternatively, we could press ahead on all fronts with the development of synthetic organisms and hope that the benefits they provide will outweigh the risks they will create.

The thing that should scare us the most about the Synthetic Age is the prospect of these types of massive decisions—literally world-shaping decisions—not being made democratically. This happens not least when the public is lured down these paths by business interests and entrepreneurs without really knowing fully what is at stake. As legal scholar Jed Purdy put it in a *Boston Review* article, we need to decide whether the world we inhabit will arise "from drift and inadvertence or from deliberate, binding choice." In order to make this choice a deliberate one, people need to know far more about the technologies heading our way.

What should be starting to come into focus is how the Synthetic Age presents all sorts of novel opportunities for earth systems management. After using nanotechnology to master the manipulation of matter and synthetic biology to master the manipulation of genomes, we likely will look around and find out what other parts of the world we inherited can be remade to suit our needs. As the idea of a synthetic world becomes more familiar, we likely will be further emboldened in our efforts to reshape the earth. We will in all probability be open to other ways of remaking the surrounding world. We will do this not just atom by atom, molecule by molecule, and genome by genome but also—as we will see next—ecosystem by ecosystem.

5 Ecosystems to Order

There was a time in the not-so-distant past when environmental conservation had an entirely straightforward focus. It was about protecting nature. The word *nature* stood for the nonhuman—the green and verdant realm that persisted independently of the influence of civilization. Nature by definition operated spontaneously and autonomously. For many, its capacity to self-organize and diversify imbued it with a significance that approached a kind of sacredness. The more independent of the works of humans nature was, the more thoroughly *natural* and valuable it appeared to be.

As humans progressively affected this surrounding matrix through their industry, nature increasingly became a different thing. This was true simply as a matter of definition. Bill McKibben, expressing a typical environmental position, explained that depriving nature of its independence is "fatal to its meaning." Nature's independence *is* its meaning. Mercury-laced halibut, climate-impacted snow packs, and radio-collared condors or grizzly bears all send clear signals of the extent of human influence. Without an independent nature, McKibben said, there is only us.

What McKibben calls the "extinction" of the idea of an independent and valuable nature is one of the most remarkable transformations of our age. It is central to the whole idea of the new, human-directed epoch. Because it eliminates a category of thing that used to put a check on human behavior, this transformation opens up all sorts of novel possibilities about how to interact with the surrounding world. To understand the significance of this change, it helps to appreciate just how deeply ensconced in environmental thought the championing of untouched nature had been.

• • •

As most of the respected historians of environmental ideas tell it, early twentieth-century environmental thinker Aldo Leopold always felt a deep stirring in his gizzard for cranes. It was not just that cranes were a spectacular species, standing four feet tall with wingspans that can cause small herds of cattle to fall into shadow. Their svelte necks, the dark sheen of their daggered bills, and the backward-flexing hinges of their fragile legs certainly make one pause and wonder at the aesthetic delights that nature can produce. But for Leopold, the significance of the cranes that migrated through the Wisconsin landscape near his home was not all about their aesthetic beauty. Nor, in fact, was it all about the cranes.

In addition to being drawn to the crane itself, Leopold also had a sense that the whole complex landscape in which cranes were found was its own object of wonder. Slow but inexorable forces had given the crane its shape and its remarkable elegance. Those same historical forces had made the marsh and its community of life into a place ripe for harboring the cranes, their food, and their foes.

Elements of raw nature such as cranes were literally the embodiment of the earth's long history. Leopold characterized the crane's call as "a trumpet in the orchestra of evolution." At the same time, he suggested that the marsh itself wore "a paleontological patent of nobility" earned through the "march of eons." A marsh without cranes, thought Leopold, was barely a marsh at all. Such impoverished marshes stand melancholic and humbled, "adrift in history."[1]

Leopold was an unusual observer of the natural world. By all accounts, he was acutely attuned to the subtle goings-on in the landscapes surrounding him. He wrote pages of detailed prose about the meander of a skunk's tracks in the snow and about the dance of a woodcock's flight in the twilight sky. But even if Leopold was an unusually careful witness, the sentiment that he tapped into when reflecting on the crane marsh has turned out to be not unusual at all. In fact, it is arguably the sentiment that has formed the bedrock of environmental thinking for the last century and a half.

Leopold is one of the foremost exponents of the idea that untouched nature is the most desirable kind of nature. The natural world and the ecosystems it contains have developed a proper shape and order bestowed on them by the long reaches of geological and evolutionary time. Untouched nature is exactly how nature *should* be. Millions of years of biological history have endowed it with a moral or even religious significance.

Leopold was not the first to advance this idea. Alexander von Humboldt, George Perkins Marsh, Henry David Thoreau, Mary Treat, John Muir, and numerous other luminaries all gestured toward this view. Humboldt was perhaps the first to see nature as a valuable "living whole" woven together "by a thousand threads." Marsh expressed admiration for how nature fashions her territory as to give it "almost unchanging permanence of form, outline, and proportion."

President Theodore Roosevelt perfectly captured the sentiment in his forthright style when he designated the Grand Canyon a national monument in 1908. "Leave it as it is," Roosevelt demanded. "You cannot improve on it. The ages have been at work on it, and man can only mar it." For all of these thinkers, nature's independence from humanity over the long reaches of evolutionary time was a large part of what made it valuable. When we interfere with these natural arrangements and the creatures they contain, we have compromised them.

In his *Sand County Almanac* Leopold calls the long-term perspective articulated by people like Roosevelt "thinking like a mountain." The land is the product of deeply rooted historical processes that originated long before humans appeared on the scene. Such antiquity demands our humility. Later on, in what became an almost sacred text for American environmentalists, Leopold delivered one of conservation's most famous lines: "A thing is right when it tends to preserve the integrity, stability, and beauty of the biotic community. It is wrong when it tends otherwise."[2]

The high moral value Leopold placed on untouched nature meant that he attributed particular importance to wild landscapes. In 1924, while working for the U.S. Forest Service, Leopold persuaded the federal government to protect a 500,000-acre expanse of land in New Mexico as wilderness, the first land given this type of protection in the nation. Since that time, the Gila Wilderness has been joined by almost 110 million acres of additional lands protected under the U.S. Wilderness Act (1964). These lands serve as a lure that pulls millions of Americans into nature on picnics, hikes, camping trips, and hunting expeditions every year. In many Americans' minds, wild lands are crucial refuges from civilization that offer the clearest window into the significance of the world outside of the human domain. The wilderness idea sparked into life by Leopold is to the rest of the world, according to some authors, America's greatest gift.

It is perhaps a wicked irony that tens of thousands of trees have been cut down to make the paper on which philosophers and environmental writers have attempted to dissect the meaning and influence of Leopold's ideas about untouched nature and the wild. Although not without its critics, this Leopoldian philosophy has been in the ascendancy, at least in North America and probably elsewhere, for the bulk of the modern environmental movement. Increasingly over the last two decades, however, it has become clear that the romantic vision of wilderness promoted by Leopold and others like him has some serious flaws. A growing chorus of dissenters has started to suggest that the environmental movement needs to move beyond ideas like naturalness and the veneration of wild landscapes toward a new vision of what environmentalism is about. Deep philosophical problems reside in the whole notion of innocuous-sounding words like *untouched*, *wild*, and *pure*. Some even say that deep moral problems reside in the very idea of nature.

There is a growing appreciation that words can be loaded in such a way that they create a distorted and unrepresentative picture of reality. Philosophers call it the *social construction of reality*. Whatever may be the true character of what lies out there, it is inevitable that people will view the world through particular cultural lenses. These lenses always lend what is seen a certain tint. Because of this tint, the words a person employs never correspond one to one with reality like reflections in a perfect mirror. The relationship is considerably more fuzzy. Often a term or a concept can say as much about the society that uses it as it does about the world it is meant to describe. Think of the different nuances of a word like *liberty* to an American, a French person, and someone from China.

If important terms and ideas can be culturally loaded, we should ask whether a phrase like *pristine nature* accurately represents the world or whether it is a distorted projection emerging from a particular sort of mindset and designed to satisfy a particular type of need. Perhaps the whole idea of untouched or virgin nature that was important to Leopold is a creation—the kind of constructed idea that fulfills one person's fantasy while being completely meaningless to another.

Increasingly, it has been argued against Leopold that only an affluent white male fleeing from an increasingly industrialized landscape and possessing unrealistic visions about the virtues of life in an earlier time would fall into the trap of seeing parts of the natural world as untouched and wild. It helps if this male belongs to an immigrant culture arriving in a land his culture has conveniently labeled *the New World.*

Leopold happened to be not only white, male, and relatively affluent but also writing in the run-up to a period known as the Great Acceleration. This unprecedented time of post–World War II economic expansion created a growing fear that America was losing the landscapes on which the parents and grandparents of Leopold and his generation had settled.

Leopold's worries about growing environmental destruction were certainly legitimate, but his idea of untouched, wild nature onto which he grafted those concerns appeared to completely ignore the indigenous presence that predated the arrival of European immigrants by millennia. It has been argued convincingly that Leopold and his followers simply failed to see the ways in which Native Americans had already transformed the landscape through game management, early settlement, fire, and agriculture. All of this had started long before the colonization of the New World by white people. European immigrants, who were accustomed to the busier landscapes of their home continent, simply looked right past the indigenous presence and settled on the idea of "wild and pristine" nature as the moral heart of environmentalism. Indigenous peoples, on the other hand—as many linguistic anthropologists note—usually have little use for the word *wilderness*. It is a term that seems to have been constructed and used only by the people who have colonialized them.

The question of whether the idea of wilderness is socially constructed is certainly both philosophically and anthropologically interesting. To some people, it connects a certain brand of environmental thinking with dark histories of colonialism and cultural genocide. But whether or not there are cultural prejudices embedded in the idea of pristine or virgin nature, to many observers the realities of the early twenty-first century mean that untouched, wild nature—if it ever did exist—simply cannot be found anymore. As advocates of the new epoch frequently observe, humans have moved mountains of earth, cleared whole continents of forest, placed dams across innumerable rivers, and built megacities across many landscapes. They have imported thousands of crops and ornamental species into some environments and have wiped out the natives—both human and nonhuman—in others. It is estimated that 39 million of the planet's nearly 50 million square miles of ice-free land has been turned over to various forms of human use.

Even in the shrinking number of places where humans have not yet set up shop, airborne and waterborne chemical pollutants have tainted every drop of ocean water and every micron of rock and soil. From Alaskan coves to the soils of the Mongolian steppe, the residues of our chemistry are everywhere. Forming a giant, aerial mantle above all this, greenhouse gases mean that all of nature's systems have to operate a degree or more Celsius above where they would be had humans not shown up.

If this is true, then it may not matter whether the idea of pristine wilderness is socially constructed. The type of environmental thinking championed by Aldo Leopold is now simply obsolete. His ideas about preserving wild lands are passé. For a number of contemporary thinkers, a new environmental movement is needed—and right on cue, it is emerging.

• • •

Emma Marris is a young science writer who has become one of the leading voices in a crusade to remake environmental thinking along different lines. For more than a decade, she has been publishing essays in magazines and journals like *Discover, Orion,* and *Nature* that promote this new vision. Raised in the Pacific Northwest, Marris and her philosopher-husband think hard about the world in which their two small children are growing up. In her reporting work on ecological issues, Marris is often as much interested in the big picture at work behind her stories as in the scientific details. Her 2011 book *Rambunctious Garden: Saving Nature in a Post-Wild World* placed her at the forefront of the hottest debate in conservation in decades. The old-style Leopoldian approach, she claims, is not just ill-informed but is proving to be a dangerous impediment to good environmental thinking.

Affable and engaging in private, Marris is helping to transform the environmental movement by being a determined advocate who argues her views with quickness and passion. Marris regularly goes toe-to-toe with environmental thinkers of the old guard like Pulitzer Prize–winner E. O. Wilson, a world-renowned biodiversity expert who is fifty years her senior. At one point in a debate with Marris, an enraged Wilson, who disapproved of her new brand of environmentalism, exploded at her: "Where do you plant the white flag that you're carrying?" Rejecting the idea that any precolonial relics of pristine nature are left to protect, Marris responded to Wilson with a line used by her friend Joseph Mascaro, "I'm here for nature, not for 1491."[3]

According to Marris, we live not only in a postwild age but in a world that is increasingly the product of countless human choices. The romantic Leopoldian ideal of preserving nature's "paleontological patents of nobility" is just that—romantic and unhelpfully anachronistic.

Marris claims that even a place like Yellowstone National Park is already so heavily regulated by watchful park managers that in some ways a "vacant lot in Detroit is wilder than Yellowstone."[4] The few remaining troops of mountain gorillas in Rwanda are now followed around at all times by armed guards to deter any would-be poachers. Simply leaving nature alone to preserve its wildness is no longer an option. Leopold's old brand of nature preservation promises an energy-sapping and ultimately futile quest. In an era in which humans are known to be causing global-scale changes, his lofty recommendation that humans should endeavor to be "plain citizens and members of the biotic community" is flawed. No species that has transformed the entire planet can be a "plain citizen" of anything.

The implication that the battle is already lost is appalling to people like Wilson. But what makes Wilson's followers really gag is where advocates of the new form of environmentalism such as Marris take it next. They suggest that if nature is gone, then environmentalism has to be less about *preserving* and more about *shaping*. No longer should governments set aside nature and protect it from any further human encroachments. It is too late for that. Humans should go out there and proactively *garden* it, both for ourselves and for the other species that share the postnatural earth with us. Environmental advocates should not withdraw from the natural world and attempt to preserve a few remaining relics of original wildness. They should manipulate it to create what is most needed, whether that is more food, better ecosystem services, or a range of spaces developed for the purposes of recreation and relaxation. This often will mean deliberately recomposing ecosystems so that they work better for us. "We are already running the whole earth," Marris asserts, "whether we admit it or not. To run it consciously and effectively, we must admit our role and even embrace it."[5]

Although those on Leopold's or Wilson's side of the debate sometimes splutter with incredulity at these suggestions, people in Marris's camp remain upbeat. The resetting of the human relationship with the environment for this new age is not supposed to be a cause for sadness. Quite the opposite. It should be thought of as a source of new and unlimited opportunities. Marris suggests we should feel excited about the possibility of meeting human needs while at the same time creating a nature that provides new possibilities for flourishing.

The new sunniness about a more self-consciously managed environment is one of the defining characteristics of the *eco-modern* thinking that is replacing Leopold's. The move to recenter environmentalism on something other than the traditional value of naturalness needs to be made, says Gaia Vince, the author of *Adventures in the Anthropocene*, without regret. "Nostalgia," Vince says, "… is a pointless sentiment." The postnatural environments we create will not be pristine or untouched, but they might have many qualities similar to those that were previously valued in what used to be called "the natural world"—only this time it will be nature version 2.0.

Erle Ellis, a geographer at the University of Maryland, mirrors the optimism exuded by Marris and Vince, stating that environmental policy now requires moving "beyond fears of transgressing natural limits and nostalgic hopes of returning to some pastoral or pristine era." Echoing Paul Crutzen, Ellis sees no option but to accept the reality that we are now "the engineers and managers of a planet transformed by the artificial ecosystems required to sustain us." We must embrace this challenge, he suggests enthusiastically, and consider our time "the beginning of a new geological epoch ripe with human-directed opportunity."[6] As many of those who champion the idea of this new epoch are quick to point out, there really is no turning back. We need to look forward, seize the future, and turn that future into the one we want most. Such thinking, according to Marris and her allies, offers us some much-needed "hope in the Age of Man."

• • •

This is heady and radical stuff for the environmental movement. It is an entirely new type of environmental thinking for a postwild planet—with nature finished, preservationism rejected, and the ghosts of Thoreau, Muir, and Leopold emphatically slayed. Clearly something big is going on. If all of this is true, then the long-standing environmental credo that nature requires protecting from the ravages of humanity has to be rejected as no longer appropriate. Humans should become comfortable in their new role as radical transformers of this *postnatural* world.

But at this point, it is worth a moment of pause. If Leopold's original environmental thinking about the value of wild and pristine nature can be faulted for the distortions created by its cultural blinders, it also is possible to point out some of the same tendencies in this rush to overturn the old environmental order. Europeans as a general rule are much more accustomed than North Americans are to the idea of a managed environment. With a few notable exceptions—including parts of the Alps, patches of the Iberian peninsula, and segments of Scandinavia—hundreds of millions of Europeans inhabit landscapes on which the transformation of pristine nature into a cultural landscape has been evident for much longer than in the so-called New World. Europeans tend not to find the idea that all of nature is already impacted quite so revelatory as do Marris, Ellis, and some of the other boosters of the new "gardening" approach to environmentalism.

Despite this acknowledgment of the extent of human influence, many Europeans remain strongly committed—even deep morally committed—to the importance of the natural world. There is also an unshaken belief in the moral significance of those predatory species that still furtively make their lives in the spaces that remain between the dense populations of human residents. This has meant a resurgent interest in charismatic and storied creatures such as wolves, bears, and jackals whose numbers, in some parts of Europe, are starting to rebound dramatically. Efforts to save and protect these highly valued components of the landscape remain vigorous.

In fact, the idea of *rewilding* certain landscapes is growing in popularity as demographic shifts taking place across Europe move human populations out of areas in which they had previously farmed. Even in a country as intensively industrialized as the United Kingdom, proposals to restore populations of wild animals such as lynx and wolves are generating significant numbers of new advocates. Beavers, wild boar, and white-tailed eagles are already back. Across the English Channel,

Germany is pursuing a national goal in which "Mother Nature is again able to develop according to her own laws" on 2 percent of its national territory by 2020.[7] The German wolf population has gone from zero to more than 250 in the last twenty years. Two of Germany's neighbors, Belgium and the Netherlands, are starting to grapple with questions about how to coexist with large predators after the recent reappearance of the wolf in their intensively managed agricultural lands. Despite high population densities, the ideas of "nature" and even "the wild" still loom large on the European radar.

The suggestion from some in North America that the world is just now entering a new age in which the fundamental orientation of environmental thinking should change toward increasing management would, to many of these Europeans, seem rather odd. The idea that environmental thinking needs to go postnatural or postwild would sound similarly bizarre. Europeans tend to accept that their landscapes are not pristine, but they remain committed to the significance of nature representing an important realm of existence beyond the cultural sphere. They are also ready to spend considerable money and time trying to enhance the various pockets of relative wildness that remain.

If these trends in Europe are anything to go by, then Leopold's idea about the moral and cultural significance of the wild may not be dead after all. In many settings, the value attached to the idea of nature operating free from human influence remains high. Wild nature may yet have the resilience to resist the probing fingers of the Plastocene.

• • •

In the light of these somewhat conflicting accounts of nature's demise, it should come as no surprise to find someone from an Old World country articulating a vision of the wild that tries to thread the needle between Leopold's and Marris's diverging conclusions. British journalist Fred Pearce offers a direction for environmentalists that recognizes the extent of human impacts across the globe but still finds room for an account of nature as lively, surprising, and wild. Using examples drawn from heavily impacted ecosystems across the world, Pearce invites a comprehensive rethinking of what exactly it is that environmentalists should be protecting. He takes a decisive step away from the legacy of Leopold but at the same time rejects Marris's notion that everything is now postwild. Pearce advocates in its place something he calls "the new wild."

As a European who has absorbed the lesson of human influence, Pearce roundly rejects the idea that only untouched ecosystems containing only native species are important for environmental thinking. Humans have been introducing species and shaping landscapes for millennia. Although people have transformed their surroundings in fundamental ways, Pearce insists that nothing about this fact prevents nature from remaining an independent and animated province.

To keep this contemporary notion of wildness alive, Pearce demands a rethinking of the widespread antipathy that traditional environmentalists have directed toward nonnative and invasive species. Blunt declarations that native species are good and nonnatives are bad are unhelpful. Such assertions are also, Pearce argues, ecologically ill-informed. Ecosystems are always a haphazard mixture of original species and new arrivals, with some of the newcomers adopting important ecological roles that those exiting from the scene have made available. Nonnative birds in Hawaii, for example, are doing most of the island's seed dispersal for its native trees. Invasive Turkey oaks in the United Kingdom have brought in a wasp whose tasty larvae have proved an essential lifeline for threatened blue tits. In Indonesia, three-quarters of the remaining orangutans live not in native forests but in tree plantations. This constant trading of ecological roles on the arrival of newcomers is not just a phenomenon of the human age, says Pearce. It is how nature has always worked. On floating logs and air currents, in the digestive tracts of birds and the fur of long-legged canids, opportunistic species are constantly on the move in search of better prospects.

Pearce claims there are sinister parallels between the prejudice against nonnative species and the prejudice found in many countries against immigrants. He argues that Nazi ideas about eugenics are linked to environmental hatred of nonnatives through similarly misinformed readings of Darwin. Despite the common belief to the contrary, it is not necessarily the fittest that survive but the most opportunistic. By illustrating the important ecological services provided by many nonnative species, Pearce shows how some human impacts on ecosystems can turn out to be for the good. By carefully investigating the rise and fall of certain immigrant species that have gone from "curse" to "ecologically desirable" in a few decades, Pearce builds the case that nonnatives have often wrongly been treated as scapegoats. They have consistently been vilified in order to take the fall for problems caused primarily by human misdeeds like pollution and habitat destruction.

A classic example of scapegoating happened in the Mediterranean Sea in the early 1990s after the algae *Caulerpa taxifolia*, brought from the Indian Ocean to brighten up home aquariums, escaped and spread rapidly across the French and Italian Rivieras. For a while, the algae appeared to be stifling native sea grasses and damaging important spawning grounds for a host of marine creatures. Panic ensued, and the algae was branded a major public enemy. Volunteers with snorkels tried to rip it up with their bare hands, with little success.

Thirty years later, however, most of the algae is gone. As soon as the urban pollution that used to spill into Mediterranean waters from beachside resorts had been cleaned up, the algae started to disappear. The health of the marine ecosystem rapidly returned. The problem, in other words, had been not the newly arriving species but people. The sea grasses had been dying from pollution even before the algae showed up. Even harder for the algae haters to accept was the fact that the *Caulerpa* arguably provided an important temporary habitat for certain native species by revegetating rocks that had been made barren by the urban runoff. The interloping algae bioremediated the pollution while providing a habitat in which native clams and cockles could multiply. As Pearce points out, invasive species are not always as apocalyptic as thought but often provide unappreciated benefits.

Pearce tells a similar story of prejudice about an infamous species in North America's Great Lakes region. When zebra mussels proliferated in Lake Erie after being accidentally introduced by a cargo ship arriving from the Caspian Sea in the 1980s, war was declared on the unwelcome immigrant. The fact that the striped invader had its origin in the Soviet Union made its arrival particularly unwelcome during the Reagan era. Dire warnings about the collapse of the lake's entire ecosystem were issued.

But zebra mussels, says Pearce, have turned out to be "the best janitors Erie ever had." They settled into a highly polluted ecosystem in which little else could live. They filtered a huge amount of contaminants from the water and have provided a reliable food source for endangered lake sturgeon, small-mouth bass, and thousands of migrating ducks that used to avoid the lake's tainted waters. Yes, the mussels have clogged pipes and created economic costs for the communities that have been forced to deal with them. They also have competed with native shrimp and clams. But the economic and ecological benefits they have provided are both underappreciated and significant. Pearce draws attention to the fact that for some strange reason, it has always seemed easier to blame the immigrant species than to look honestly at our own failings.

To further illustrate the muddy thinking, Pearce points out that exotic arrivals do not just occasionally turn out to be ecologically valuable. They sometimes become idolized as heroic and welcome residents. Twelve American states—from Nebraska to New Jersey—have designated the nonnative European honeybee as their official state insect.

These cheerful interlopers are cherished in part because they now perform 80 percent of U.S. crop pollination. In addition to those living wild, tens of millions of European honeybees in thousands of traveling hives are trucked across the nation, from California to New England, in a carefully timed vehicular migration to pollinate key agricultural crops as they blossom. These economically important plants include the apple, blackberry, blueberry, cantaloupe, cherry, clover, cranberry, cucumber, eggplant, grape, lima bean, okra, peach, pear, pepper, persimmon, plum, pumpkin, raspberry, soybean, squash, strawberry, and watermelon. The helpful foreigner ends up supporting large sections of the U.S. agricultural economy through its free labor. The U.S. almond industry, for example, is entirely dependent on the nonnative honeybees for its existence. In California alone, this industry supports 104,000 jobs and adds more than $11 billion in value to the state's economy.

Even without such unquestionable ecological and economic benefits, Pearce points out, there is a more pragmatic side to the argument in favor of nonnatives that comes from the very character of today's landscapes. The sheer extent of introduced species now present in almost every landscape suggests that there can be no turning back. Thirty-five percent of the species in the San Francisco Bay estuary and a quarter of the species in the Florida Everglades are nonnatives. There are more camels now living in Australia than in Saudi Arabia. On islands like Hawaii, nonnatives make up more than half of the flora and fauna. Removing them one by one would be impossible, and it is far from clear whether what was left would function in any desirable way.

Within agriculture, the presence of nonnatives is even more striking. Nonnative species make up nearly 70 percent of food crops worldwide, a number that rises to 90 percent in the United States and nearly 100 percent in island nations like Australia and New Zealand. Domesticated animal species (sheep, cattle, pigs), often introduced, are dominant. A full 95 percent of the earth's terrestrial biomass is now made up of the combined weight of humans and domesticated agricultural animals. These endless mountains of nonnative flesh that crowd farms and concentrated animal feeding operations around the world make it likely that introduced species are here to stay. And the continuous, global species exchange is unlikely to end anytime soon. At every moment, there are between seven and ten thousand species traveling to new destinations in the ballast water of the world's cargo ships. Biologists talk of how the international movement of plants and animals has effectively recreated Pangaea, the single supercontinent that existed until about 200 million years ago and for which ocean barriers to migration were irrelevant. When talking about nonnative species invasions today, it is too late to talk about shutting the barn door. The nonnative horses have already bolted.

Human influence on nature is dramatic and real, but like Marris and others who embrace a human-impacted ecology, Pearce insists that this is no cause for regret. Invading species have always propelled nature forward. Pearce documents how even abandoned industrial sites can become new hives of biodiversity. He talks enthusiastically about slag heaps from coal-fired power plants as "brilliant oases of biodiversity" and about a heap of pulverized ash at an abandoned Thames estuary site as "a treasure trove" of orchids and invertebrate species. For Pearce, introduced and immigrant species are the key to keeping biodiversity high and ecosystems functioning smoothly.

The new wild that Pearce champions has very little to do with preserving historical remnants of nature in a state that is as pristine as possible. It is about allowing and sometimes facilitating ecological change. By the end of the Holocene, a large number of ecosystems across the globe have already become irreparably *novel*. This new term of art in ecology refers to any ecosystem has been deeply influenced by humans, contains species arrangements that have never been present before, and is unlikely to be budged out of that new state.

What is new for environmentalism is the idea that this human influence is not something that needs to be rooted out. Healthy ecological processes, Pearce insists, are driven by the opportunism that produces novel ecosystems. This opportunism happens on the back of human disturbances. Pearce even whispers the heretical view that this positive account of the new wild might spill over into how we think about climate change. With warming temperatures, an explosion of evolutionary activity might be about to take place. Species are moving, hybridizing, and developing innovative survival strategies. For Pearce, this innovation is not something to lament but an illustration of nature operating at its best: novel ecosystems are the welcome heart of the new wild.

Pearce's articulation of a new wild is significant for what it says about how humans should interact with nature. It turns traditional environmental thinking more or less on its head. After you have surrendered the goal of trying to preserve ecosystems in some sort of historically favored state and have accepted the reality of constant change, a number of new possibilities for landscape management open up. Within some of these possibilities, Pearce's vision of the new wild and Marris's idea of the postwild show signs of converging.

If the presence of introduced species in an ecosystem is not something to lament, then the deliberate shuffling of an ecosystem's constituents may not be as unacceptable as environmentalists had traditionally thought. In order to protect some of the species that we really value, we should not build fences, keep out other species, and try to preserve some fixed state from the past. We should proactively intervene in the natural order by moving and exchanging species in order to recompose ecosystems intelligently and deliberately. We should not be shy about cutting and planting, importing and hybridizing, introducing and reworking that land around us.

Nature, in other words, might paradoxically need considerable human manipulation in order to survive *as nature* in the new epoch. In Marris's terms, we must start thoughtfully "gardening" the world around us. And this does not just mean gardening the cultivated spaces near to our towns where we produce our food and keep our animals. It means gardening the whole thing. The entirety of nature may now be our farm.

• • •

It is a fundamental tenet of many religions and the belief systems of many indigenous peoples that we are born into a cosmos created by a power other than us. Creator stories serve the purpose of pushing explanations of origins in the direction of a large and spiritually significant force. As a result of these explanatory tales, many traditions have felt called on to treat their surroundings with respect due to their sacred origins. Although this respect can manifest itself in many different ways, the holy origin of the surrounding world has put constraints on how humans are supposed to act toward it.

Even those who do not subscribe to any sort of divine explanation for the earth's origin often marvel at how the physical and chemical forces responsible for creating the world predate the arrival of humans by billions of years. Paleontologist and evolutionary biologist Stephen Jay Gould captured this unfathomably long period prior to the arrival of *Homo sapiens* with characteristic aplomb: "Consider the Earth's history as the old measure of the English yard, the distance from the king's nose to the tip of his outstretched hand. One stroke of a nail file on his middle finger erases human history."[8] The king's outstretched arm represents an expanse of time in which something clearly remarkable took place, with all of it occurring completely independently of human interference. This is the paleontological patent of nobility to which Leopold referred when he spoke so favorably of the crane. In the eyes of many environmental advocates, the workings of the natural world over this long passage of time deserve our respect.

The novel type of approach to ecosystem management now being advanced by Marris and other new conservationists therefore represents a radical turn for environmental thinking. A former governor of Alaska not known for his environmental sensibilities was widely ridiculed when he defended his plans to shoot wolves by stating, "You can't just let nature run wild, you know." The ideas being proposed by some of today's new conservationists offer an updated and more informed version of this governor's statement. For them, nature does not need to be left to its own devices; it needs to be shaped. Just as it was with synthetic biology and DNA, nature need not be preserved in its historic form but should be reconstructed along better lines. The Synthetic Age presents an opportunity for humans to dramatically improve the biological and ecological world they inherited.

In this new epoch, nature protection starts to mean something entirely different. This is a type of thinking that takes conservation in directions that would shake Aldo Leopold to his wilderness-preserving core.

6 Relocating and Resurrecting Species

When the old idea of nature's pristine harmonies is abandoned, the door is opened for a dramatically more interventionist type of environmentalism. Humans have already been involved in haphazardly shaping nature. They might as well become more thoughtful and deliberate about it. Paul Crutzen suggested that it is up to humans to decide what nature is and shall be. Some ecologists are clearly excited to embrace this opportunity.

As the impacts of anthropogenic climate change have become more pronounced, it is becoming clear that many species will simply be cooked if they remain in their historic geographical ranges. In the United Kingdom, for example, lines demarcating average annual temperatures are shifting north at a rate of just under three miles per year. For some organisms experiencing climate stress, it is relatively easy to pack up and move to where the climate is more suitable. If you have wings or muscled legs and a fairly flexible diet—think of a magpie or a fox—you might be able to move your range those few miles north each year without much trouble. But if you have roots or an isolated hillside on which you live or simply do not like to do things in a hurry, migrating at the required rate may not be an option. Most trees, for example, are unable to move north through seed dispersal at more than one hundred feet per year. It is even worse for earthworms. In some circumstances, these humus lovers are said to expand their range at little more than a mile each century. Species like these simply will not be able to outrun climate change.

Biologists emboldened by the new environmentalism are increasingly suggesting that struggling species should be given a helping hand. *If*, as Fred Pearce has documented, species locations are already significantly shaped by human interventions and *if* this is morally and ecologically acceptable, then it might not be a big deal to proactively relocate vulnerable species into areas where they stand a better chance of surviving. Assisted migration—rebranded by some of its advocates as *managed relocation*, given the increasingly politically charged connotations of the word *migrant*—is a new technique for dealing with climate change that has many traditional environmentalists all tied up in knots.[1]

• • •

Biologist Chris Thomas from the University of York exudes an energy that at times is remarkably similar to some of the insects he studies. A lean man with steel-rimmed glasses and a closely shaven head, Thomas lights up when the topic turns to butterflies. Like many successful academics, his passion for his subject matter utterly consumes him. A recently elected fellow to the U.K.'s Royal Society, Thomas's research interests orbit around ecological and evolutionary responses to the effects of climate change, including habitat fragmentation and species invasions. He is particularly concerned about how the warming climate will affect birds, plants, and insects, and he has tried to determine the conservation strategies that might be necessary for saving them.

But beware of trying to have a leisurely conversation with Thomas on a summer afternoon. In the midst of the conversation, his eyes might start to trace the flight of a passing *Lepidoptera*, his head weaving up and down gently as he follows the insect's path through the sky. Before long, you realize that he is no longer listening to you, at which point you might look around to try to spot the insect that has caught his attention. If you provide half an opening, Thomas will be gone, taking off across the field with his long arms and legs whirring like a locust in pursuit of his quarry before he bends down and takes off his glasses to study from inches away the specimen that has caught his attention.

A few years ago, Thomas and a pair of colleagues, Jane Hill and Steven Willis, embarked on one of the earliest experiments in managed relocation.[2] Concerned about what climate change was doing to the prospects of a pair of local butterfly species—the marbled white and the small skipper—the research team decided to try something special. They loaded up a couple of boxes containing about five hundred individuals of each species in the back of a car and jumped on the motorway heading north.

Butterflies, you might think, ought to be capable of simply flying themselves out of trouble if their native range starts to heat up. But this is not always the case. Barriers to dispersal, such as large urban areas or a stretch of intervening habitat with the wrong food supply, can sometimes make a self-powered relocation impossible. Some butterfly species are also homebodies and simply do not like to travel. A combination of these factors made these two species of *Lepidoptera* particularly unlikely to beat out the rising heat on their own.

After a quick trip up the A-1 highway in northeastern England, the butterflies were released into two quarries that Hill, Willis, and Thomas had determined were suitable habitat. Because the quarries were essentially abandoned industrial sites, there was no concern that the migrants were going to disrupt any sort of pristine natural order. Local conservation experts served as advisers, and butterflies of both sexes were released at the same place in each quarry within hours of being netted from their habitat farther south. Then they were left alone to do their butterfly thing.

Careful studies in the decade since the relocation have indicated that not only have the two species survived in their northerly home but the populations have grown and dispersed in the way that a happy population of butterflies would be expected to do. The managed relocation worked and worked well. It was cheap, apparently benign, and effective.

From Thomas's point of view, this small test showed that managed relocation has promise as a conservation tool for limiting the impacts of climate change on slow-moving species. If it works for butterflies, Thomas reasons, it could work for other species. It provides hope that species threatened by climate change can be given the assistance they need to survive and is an example, Thomas would argue, of climate-smart conservation.

Unfortunately, the rates of dispersal the researchers have found at the new site—which fairly closely matched the typical dispersal rates of other butterfly species—are substantially less than the distance at which British temperature lines are moving northward. On the one hand, slow dispersal means that the butterflies are not going to spread like a plague and become a menace in their new home. But on the other, it also suggests that butterflies and similar species may need to be moved multiple times over the coming decades if they are to keep up with climate change. Intervention will increasingly become the norm.

• • •

Even if it can work for certain species, many biologists and environmentalists are deeply uncomfortable with the whole idea of managed relocation. They wonder how you can know that a relocated species will successfully adapt to its new home and whether you can be sure the introduced species will not wreak its own biological havoc. Animals outside of zoos are called "wild" for a reason.

Some relocated species, including red wolves in Great Smoky Mountain National Park and the first few Canadian lynx reintroduced in Colorado, starved to death after being released in what experts had assumed would be excellent habitat. Others, such as the European starlings that were let loose in New York City's Central Park in the 1890s as a rather unusual commemoration of their mention in Shakespeare's *Henry IV*, fared considerably better. The starling has spread across the continent and now numbers over 200 million individuals, making it quite likely the most numerous bird in North America.

The risks to the welfare of the released animals and to the surrounding ecology posed by intentionally relocated species have earned the practice the label of *ecological roulette* from those who remain skeptical of nonnative species. Whether this reflects the accumulated prejudice against immigrants that Pearce documented or the risks built into the practice of intentionally shuffling species is still being debated. But there is an inherent unpredictability about what will happen when species are deliberately moved into habitats they did not colonize themselves.

Deeper philosophical puzzles also arise. Does the relocated marbled white butterfly still wear that "paleontological patent of nobility" in its new home that Leopold had suggested was "won in the march of eons"? A relocated marbled white has certainly not marched with the eons under its own steam. It has been driven up the A-1 motorway by Chris Thomas in his Ford Fiesta. Whether you think this human intervention has somehow sullied the integrity of the butterfly probably depends on whether you think nature remains "nature" after humans have started intentionally reordering it.

The level of intervention on display in assisted migration takes us beyond the largely accidental and haphazard shuffling of species that has taken place throughout human history. It also takes us beyond the realm of moving species for pleasure or for economic gain. It begins a new practice of moving species ostensibly for their own good, with their good being determined by benevolent and well-informed wildlife biologists. However well-intentioned and well-informed these biologists are, these ultimately will be cultural choices made about which species should be moved. Managed relocation means that the species composition in a particular ecosystem is determined by humans rather than by nature.

For many people, this goes against their basic understanding of what nature is. One environmental philosopher has suggested that relying on human choices about what the natural world should look like amounts to "faking nature."[3] He doubts that a human-designed ecosystem populated by species that we have chosen to put there deliberately remains a natural ecosystem at all. Nature's independence, as Bill McKibben pointed out, is essential to its meaning.

If introducing an entirely new species to an unsuspecting ecosystem sounds like too much meddling, another type of managed relocation walks a still finer line. Less dramatic than what Chris Thomas did with the five hundred marbled white butterflies is the idea of importing specially selected individuals of a species possessing certain valuable traits into a damaged ecosystem.

Whitebark pine is a tree species that lives in high-elevation montane locations in North America. It is suffering badly at the hands of blister rust and pine bark beetles, both of which have recently become much more threatening to the pine as a result of climate change. In addition to the ghostly skeletons of thousands of ancient whitebarks now littering the high-elevation landscapes, the younger pines are succumbing to pests and diseases before they can reach reproductive age.

Quite apart from being a beautiful and tenacious tree, the whitebark pine plays a key role in Rocky Mountain ecosystems. A distinctive environment has evolved around the species over many millennia. A jaylike bird called the Clark's nutcracker disperses the seeds of the pine cones. Grizzly bears have learned to scarf down as many of these high-energy seeds as the Clark's nutcracker does not get to first. Numerous other species sensitive to the rate of spring snowmelt in the high country, from scavenging wolverines to delicate mosses, have their destinies tightly woven into that of the pine. Already the decrease in whitebark pine seeds available in early fall is thought to be forcing bears to forage in different habitats, in the process increasing their likelihood of bumping into humans. Without the pines to provide shade, the high-elevation landscape dries out more quickly as the snows recede faster in the spring, leading to a cascade of knock-on effects.

Refusing to leave the pines to their fate at the hands of climate change, a botanist at Crater Lake National Park named Jennifer Beck has organized the transplanting of seedlings descended from particular whitebarks growing in the park that appear to be more disease-resistant than others. These genetically advantaged whitebarks are tested for resistance and then propagated in an off-site nursery for a number of years before being taken back up to the ancient caldera and planted alongside their ailing cousins. Beck hopes these transplants will give the natural high-elevation stands more of a chance.

This interventionist strategy stops short of moving a whole species to an entirely new location. The transplanted whitebarks are merely a genetic variation of a species that is already there. Yet the procedure still involves a heavy dose of intervention. It means that humans are making decisions that over time will shape the genetic composition of the natural ecosystem. It is a genetic remaking of nature according to what we hope will work better. There is a genuine species-to-species altruism involved. It is not about choosing what is best for us and our pocketbooks. It is about choosing what appears to be best for the pines. Nevertheless, a system that had always composed itself without anybody's help is now being recomposed by human gardeners. Nature is no longer left alone.

Similar strategies (known as *assisted evolution* or *facilitated adaptation*) are being pursued in the Seychelles with the breeding of heat-resistant corals that can better survive the stress of higher ocean temperatures. If these colonies are successful, they will be transplanted to the sites of reefs in precipitous decline due to the consequences of climate change. In the U.S. northeast, chestnut trees are being bred to be resistant to the blight that decimated the mast forests of New England in the early twentieth century. Given enough time, the coral, the whitebark pine, and perhaps the chestnut would have most likely evolved their own resistance to the conditions that are harming them. But with the rapidity of climate change, time is in short supply for many species. Humans have therefore decided it is necessary to intervene. In these situations, evolutionary and ecosystem processes are no longer working entirely independently of us.

Environmentalists of the Leopoldian school are suspicious of this kind of intervention. At forested locations in Washington state, they have sought to prevent any transplanting of trees from occurring inside areas that are designated wilderness. In wilderness, they argue, the land is intended to be entirely self-regulating, even when the proposed interventions are for the purpose of saving an iconic species. For those wilderness advocates still deeply committed to the familiar notion of naturalness, letting nature be independent of human manipulation turns out to be a higher priority than any one species' survival. Plus, risks are involved. Intentional tweaking of the ecosystem, they suggest, not only destroys its inherent wildness but will almost certainly lead to unexpected consequences. Biological nature contains too many unknowns. Our science is too inexact. Our interventions are too clumsy.

In some cases, they may be right. Humans have a sorry record when it comes to anticipating the outcome of transplanting species beyond their native range. Kudzu in the American South, European rabbits in Australia, and water hyacinth in Africa's Lake Victoria did not turn out particularly well for the local ecosystem. Unanticipated and expensive actions to dial back the harm often have to be undertaken.[4]

Jennifer Beck and her crew at Crater Lake have already found themselves more involved than they anticipated. They have had to go into the landscape and "strangle" encroaching mountain hemlock trees in order to give the whitebark seedlings a chance. This involves cutting the hemlock's bark all the way around the circumference in order to destroy the tree's ability to transport nutrients—killing a native species to save a transplanted one. The ethics seem to become more twisted the more humans get involved.

While ethicists ponder the morality of assisted migration and assisted evolution, the commercial world is not standing by idly. New technologies for shaping evolution are breaking through at a startling pace. Jennifer Beck's strategy of searching out disease-resistant strains of whitebark pine on the mountainside and then propagating them in nurseries is starting to become distinctly old school.

• • •

CRISPR is an acronym that stands for Clustered Regularly Interspaced Short Palindromic Repeats. It is part of the defense system employed by bacteria against harmful viruses. Bacteria that have survived a previous viral infection can store short sequences of the hostile DNA as a type of "biochemical memory" of their enemy. When that enemy invades again, the bacterium is able to identify it, bind to the dangerous portion of the DNA, and cut it out. This renders the invader harmless. The bacteria also can replace the piece of the genome that has been cut with a different, more desirable gene sequence.

Scientists in Japan, the Netherlands, and Spain identified this bacterial mechanism more or less simultaneously in the late 1990s. After a decade of incremental gains in understanding the biochemistry involved, a Lithuanian researcher named Virginijus Siksnys showed that the "gene-editing" mechanism could be transferred to other bacteria. In 2013, researchers at Harvard and MIT worked out how to use this finding in the genomes of more complex organisms beyond simply bacteria. This made available what was essentially a highly efficient gene-editing technique for use on plants, insects, and even mammals.[5] CRISPR meant that genomes could be cut in precise locations and the material removed could be replaced with gene sequences selected to perform useful functions. An agricultural crop, for example, could have its genome edited to resist a blight. Diseases with identifiable genetic causes could be targeted so that the harmful DNA is removed. A modification of the CRISPR technology, rather than taking genes out, allows genes to be turned on and off or be stimulated and muted so that they can express themselves in ways that can be regulated.

Taken together, these developments mean that the days in which Jennifer Beck has to hike up steep mountainsides searching for disease-resistant whitebarks may be coming to a close. If a gene (or set of genes) that enables pines to resist the blister rust is identified, CRISPR techniques may make it possible to insert those valuable genes directly into the germline of pine trees so that a better tree can be grown in laboratory conditions. No one will need to haul sacks full of pine cones back to the lab after tiring and time-consuming field trips. Researchers could stay home and use genomic technologies to manipulate existing pines in the lab.

The giant leap forward made possible by precision gene editing has the potential to put all sorts of potentially life-saving modifications of besieged organisms on the table. Bull trout struggling to adapt to elevated temperatures in high-altitude mountain streams could potentially have a gene for heat tolerance inserted into them. Highly endangered black-footed ferrets suffering from generations of inbreeding could have their genetic diversity increased by the insertion of genes from specimens in museums and frozen repositories. They also could be engineered to resist the sylvatic plague that threatens both them and their prairie dog prey. Honeybees subject to colony collapse disorder could be genetically enhanced by the addition of genes for the fastidious hygiene traits found in some colonies that have proven successful in keeping hives free from parasites. Bats with white-nose syndrome, amphibians with chytrid fungus, and Tasmanian devils with facial tumor disease could all theoretically have beneficial genes inserted through CRISPR. Conservationists might have a technology to support their own Christmas wish list.

Although gene editing can work on only one genome at a time, a new technology called a *gene drive* enables fast-breeding populations to spread engineered traits quickly through their wild populations. One version of a gene drive puts the CRISPR editing mechanism primed with the desired trait into the germ cells of a reproducing organism. If a doctored organism mates with an individual that lacks the beneficial trait, the CRISPR technology—now embedded in the germ cell—will edit the replacement trait into the chromosome that lacked it. The new individual now has the valuable trait present in both chromosomes and is ready to pass it on to the next generation, greatly improving on the 50 percent chance of inheritance that otherwise would have been present. Also passed on is the still functional CRISPR editing mechanism. The editing process and the valuable gene now will spread quickly through the wild population as it continues to breed with an almost 100 percent chance of passing on the desired genes.

CRISPR and gene drives dangle the carrot of genetic modification operating for the first time beyond the agricultural and domesticated realm. Humans could potentially change the genetic make-up of animals that never make it anywhere near a lab. The quicker the wild organism reproduces, the faster a gene drive will allow a trait to spread throughout a wild population. Most large mammals are poor candidates for gene drives because of the long time they take to reach reproductive age. Insects, on the other hand, have much more promise. In one project motivated by strong humanitarian motives, scientists are trying to work out how gene drives might be used to create populations of mosquitoes incapable of carrying the parasite that causes malaria. Gene drives promise the capacity to manipulate wild nature directly. It is, as one lab at MIT devoted to these sorts of projects claims, an opportunity for "sculpting evolution."

• • •

The nineteenth-century English political philosopher John Stuart Mill once suggested two different ways of thinking about the word *nature*. One is to suppose that the term designates everything that happens on the earth that is consistent with the laws of nature—in other words, everything that is not supernatural. With this meaning, bears, waterfalls, whitebark pine seedlings bred in a nursery, Park Service employees killing hemlock trees, and CRISPRed mosquitoes are all part of nature. They do not transcend any physical laws. To transcend physics, one has to be either an angel or a god.

Mill's other way of thinking about nature supposes that nature includes everything that takes place on earth *with the exception of* what happens as the result of human intervention. In the first case, humans and all their works are entirely natural. In the second, nothing about humans or what they do can be thought of as natural. In this second case, every house, automobile, and vegetable garden is unnatural. Synthetic organisms are unnatural. By this definition, the activities of Jennifer Beck at Crater Lake and those of the gene-editing scientist are unnatural. Mill's distinction captures two polar opposite views of humanity. The first puts them entirely within nature. The second puts them entirely apart from nature.

Chris Thomas and Jennifer Beck might not care much about John Stuart Mill, but it would certainly boost their philosophical position if they both adopted Mill's first view of what counts as the natural. In this case, calculated human interventions—such as putting a box of butterflies in a car and driving them north or growing rust-resistant whitebark pines in a nursery before transporting them to a mountainside—do not negatively affect the butterflies' or the pines' naturalness. They were natural in their home range, and they remain natural after their human-assisted relocation to the new range.

The argument has some sense to it. Humans are a product of Darwinian evolution. Why should our actions stand out from those of the rest of nature? As biological beings, we are simply exploiting some of the capacities and talents that evolution blessed us with. Nothing unnatural there. This seems especially true when the motivation behind the action is to save a piece of nature from extinction rather than to exploit or destroy it.

Tempting as this sweeping account of naturalness is, it comes with certain costs. Such a position makes it possible to suggest that absolutely anything humans do is natural. Cutting down and paving over forests? Natural. Tossing empty beer cans into the creek? Natural. Creating toxic waste dumps? Extremely natural. Raising the planet's temperature and extinguishing gazillions of species? Oh so very natural. The all-encompassing first option Mill offers robs us of the ability to condemn any human behaviors on the grounds that they are unnatural. Human actions are always part of nature by definition.

The opposing view was adopted by Bill McKibben when he suggested that nature's independence from humans defines it. The natural world by definition has to be the world unmodified by humans. After nature's independence from humans goes, naturalness vanishes with it.

The problem with McKibben's position is that because the influence of humans is now so widespread across the planet, this second version of the natural no longer seems to apply to anything left on earth. Given how comprehensive human influence is today, from thoughtlessly spewed mercury to widely emitted greenhouse gases, holding out for pristine nature appears to be a waste of time. It is pretty clear that we have left the Holocene, so now we need to talk about how best to humanize the planet rather than imagine it in a nonhumanized state that is no longer possible. The interventionist ideas being put forward by new pioneers like Emma Marris, Fred Pearce, and Chris Thomas all reflect this belief. Waxing lyrical in a Leopoldian vein about the value of untouched nature starts to appear increasingly detached from reality.

Traditional environmentalists like E. O. Wilson are incensed by the idea that humans can give themselves the moral authority to tinker with every ecosystem. Rallying around McKibben's call for restraint, they ask if we might find within ourselves the humility to leave some places alone. Haven't we destroyed enough already?

But Marris counters that to think of humans as so separate from nature that their slightest touch irredeemably taints the land is itself a sign of arrogance rather than humility. As a species that emerged from the same evolutionary processes that have shaped the natural world, Marris thinks we are simply not very different or special. In a world of ecological hurt, we need to be prepared to intervene on behalf of those species we want to save. If we sit back and leave nature to its fate, Marris suggests, we will do so with "blood on our hands." Intentional ecosystem engineering is not just doing what is practical. It is doing what is ethically required.

Marris admits that she sometimes feels awkward about such a highly intrusive philosophy because it goes against what remain some strongly held intuitions: "We have pulled a few species back from the brink—the California condor, the Whooping crane—by insinuating ourselves in their lives as puppet mothers and migration guides, so intimately that I squirm at their lost dignity and wildness." Yet she does not stop her reflection there: "But then I remind myself: that dignity trip is my baggage not theirs. They just want to live."[6]

If we care about the species threatened by climate change and other harmful anthropogenic impacts, we need to be prepared to stitch together ecosystems that will work for more of our favorite creatures within them. That, say those who side with Marris, is the essence of a climate-smart approach to conservation.

The idea that acceptable environmental management today involves decision making that ramps up, rather than withdraws from, human interference with nature completely changes the conservation game. With "hands off" no longer the preferred option, the idea of protecting nature from human influence so that its sheer independence from us can inspire us with awe becomes moot. We simply have to bite that bullet.

A growing chorus of self-styled ecomodernists insist that this is not such a bad place to find ourselves. There will be plenty of room for awe in a "new nature" that is crafted by human designers. Nature can still be sovereign and creative, as Pearce found out while becoming more and more enthralled by the "new wild." And Pearce is not capturing even half of what is on offer. Some other enterprising molecular biologists are pretty sure they can guarantee us something awe-inspiring by taking us far beyond the challenge of saving existing species. These are the biologists who plan to bring back extinct species; the people who think that before too long we may once again find ourselves standing nose-to-nose with woolly mammoths.

• • •

Located at an extreme end of the interventionist spectrum, deextinctionists—or *extinction reversalists*—embrace the possibility of not just reorganizing ecosystems by moving species around but of recreating extinct species so that lost biodiversity can be regained. It turns out that the same techniques now available in synthetic biology for building genomes can be put to use reconstructing the DNA of extinct animals. Extinction, these biologists propose, need not be forever after all.

To pull off this type of Lazarus project, all you need is a copy of the DNA of the extinct species. And it just so happens that some extinct species, such as the passenger pigeon and the Pyrenean ibex, disappeared recently enough that fragments of their tissue were intentionally preserved for scientific interest.[7]

If you are lucky enough to have a complete genome at hand, you could theoretically transfer the whole thing into an egg cell of a close living relative that has had its own DNA removed. This close relative might be a domestic goat for a Pyrenean ibex or an Indian elephant for a woolly mammoth. The technology for doing this, known as *somatic cell nuclear transfer*, has already been developed and employed successfully to clone sheep, cats, deer, oxen, rabbits, horses, and dogs. You would essentially be using a well-established cloning technique to create a clone of an extinct animal using an existing animal's ovum.

When the egg cell with its inserted DNA has divided a few times, it can be implanted in the womb of the relative species so that normal gestation can proceed. If the resulting embryo survives the pregnancy period inside its surrogate mother, then something very close to the extinct species will be the result. An "extinct" Pyrenean ibex was born using this technique to a goat mother in 2003. Unfortunately, the resurrected ibex survived only ten minutes outside the womb because of the presence of major defects in its lungs.[8]

In cases where you do not have a copy of the complete genome of the extinct animal, you might be able to figure out what it looked like. With the woolly mammoth and other extinct mammals like the cave bear, a considerable amount of fragmented DNA can be extracted from remains recovered from the permafrost or deep inside caves. Careful comparisons to the genomes of closely related living species permits evolutionary biologists to work out something very close to the extinct animal's genome. Even though it contains a massive 4.7 billion base pairs, a blueprint of the genome of the woolly mammoth is already available.

Because whole genomes of birds and mammals are very long—much longer than those of the yeasts and bacteria that Craig Venter has been working with—the best strategy for synthesizing such a genome is to start with a close relative and to use CRISPR gene-editing technologies to replace the most defining sections of the "living animal" genome with the equivalent sections of the "extinct animal" genome. You could, for example, insert the genes that code for the woolly mammoth's curved tusks into an Asian elephant genome. You then could insert the genes for the hairy skin of the woolly and the genes for the hump on its back and for its thermo-regulation in cold climates. The Asian elephant's genome still would form the core of the in-process woolly mammoth genome, but as more and more edits were completed, it would look progressively more and more like the genome of its extinct cousin.

Birds present an additional challenge because their embryonic development takes place within the yolk of a hard-shelled egg that is constantly on the move down an oviduct. It would be very difficult to replace a natural embryo with a cloned one when the former is only a few hundred cells big, is surrounded by a yolk, and is continually in motion. A promising alternative being developed for extinct birds like the passenger pigeon is to engineer the DNA of cells that would be found in the reproductive organs of the extinct species and inject them into embryos of the living species. The injected cells would naturally migrate into the gonads and start reproducing there. This would result in individuals of a living bird species whose gonads are busily multiplying the germ cells of an extinct bird species.

Such an unusually altered individual would be a chimera. Like the creature of ancient Greek mythology with a lion's head and a goat's body, this chimerical bird would be a mix of two species. Most of its DNA would be from the living species, but the DNA it passes on to the next generation would be that of the extinct species. When two of these chimeras breed, they would produce something that is pretty close to a living version of the extinct species. The result would be a creature whose entire genome is of the extinct bird, even though the newborn's parents were both of a different species. Just as happened with the creation of a synthetic bacterium described in a previous chapter, the use of synthetic biology for deextinction would have put a human twist on Darwinian principles of descent.

Whether you are using the whole genome transfer method for mammals or the chimera method for birds, the resulting creature will not be a perfect example of the extinct species. Other factors come into play. The resurrected ibex born in 2003, for example, was not a pure Pyrenean ibex because it had been incubated by a different species, resulting in embryonic development that combined factors from both its extinct animal genome and its surrogate parents. A woolly mammoth embryo would quite possibly pick up some non-mammoth DNA *in utero* from its Indian elephant host—a phenomenon known as microchimerism—resulting in a slightly less than woolly mammoth infant being born.

Nurture also would play a confounding role. As soon as the woolly mammoth was born, its Indian elephant parents would start raising it in an Indian elephant way. The youngster would be the strangest sort of hybrid—an individual living at the intersection of two species. Its genetics would be largely those of the extinct species and its natal environment would belong to an existing species. The newborn would respond to both of these factors.

These quibbles about a slightly confused identity tend not to bother the deextinctionists. In this post-Leopoldian era of human-induced change, the obsession in conservation with historical accuracy and natural purity has already been relaxed. As one of deextinction's most vocal advocates, founder of the *Whole Earth Catalog* and environmental entrepreneur Stewart Brand, glosses it, "The results won't be perfect … but it should be perfect enough. Nature doesn't do perfect either."[9]

In subsequent generations, as knowledge of the most defining parts of the extinct animal's genome became more available, the germ cells of these imperfect individuals could be systematically tweaked so that the species becomes more and more like the target animal or bird. Traditional back-breeding in which individuals with the desirable traits are deliberately crossed with each other could also help. With improvement in the technology and further concentrating of the extinct species genes in future generations, something progressively closer to the missing species could be recreated over the course of several generations. Brand notes that a generation for a woolly mammoth is about twenty years, so this means that, to deextinct a woolly mammoth, researchers need to be committed to a project that will take a century or more. But if such an investment of time and resources is deemed worthy, selected species killed off in the megafaunal extinctions at the end of the Pleistocene once again could roam suitable habitats, increasing biodiversity and assuaging a small portion of the guilt that humanity carries around for killing them off in the first place. As Brand sees it, deextinction could serve as a source of atonement for some of our previous ecological sins.

Whatever the prospects and time horizons for the technologies themselves, ethical questions about deextinction are already generating a great deal of debate. As the experience of the ibex shows, cross-species clones almost always contain genetic defects. Deextinction attempts would almost certainly cause suffering for the individual animals created as the kinks in the technology are worked out. They also would create unpleasant and probably harmful experiences for the surrogate parents who find themselves raising a newborn they do not quite recognize. "Elephants do not fare well in captivity," says Beth Shapiro, author of a book on the technologies for creating wooly mammoths: "They struggle with assisted reproduction and should be allowed to make more elephants."[10]

Furthermore, the first few individual wooly mammoths born would be some of the loneliest creatures imaginable, cut off from their kind by the intervening millennia. Viewed through a lens more sympathetic to the animals involved, switching out the genome of a fertilized egg in a mammalian species and replacing it with something created in a lab can start to look less like a moment of redemption for our species and more like an unpleasant form of genetic hijacking.

Ecologists also are concerned about the rolling of the dice involved in putting a missing species back into an ecosystem that may no longer be able to support it. This is a close parallel to the worry known as *ecological roulette* that haunts the practice of assisted migration. Who knows what sort of ecological consequences might follow from the reintroduction of an extinct species? Although the Pyrenean ibex would go back into an ecosystem it only recently left, the case of the cave bear or woolly mammoth is different. Each would now essentially be an alien species in an environment substantially transformed from its past conditions, not least by the inexorable forces of human-caused climate change.

To hedge against the question of suitable habitat, Russian ecologist Sergey Zimov is already preparing a "Pleistocene Park" in Siberia for the return of the mammoth. Zimov is grazing the mossy and forested tundra with herbivores like musk oxen, reindeer, and Yakutian horses in the hope of returning the landscape to a grassland steppe. This is the ecosystem that was in place when the woolly mammoth was still around and busy shaping it. Zimov's long-range vision involves the restoration of a Pleistocene landscape that will be a suitable habitat in which to place the first herd of deextincted woolly mammoths, ready perhaps a century from now.

Even if the habitat was available for the deextincted species, it is unclear what other kinds of price today's struggling species would have to pay. Conservationists on thin budgets are worried about the financial costs of deextinction and the ways that the technology could divert attention toward a handful of flashy animals at the expense of the less glamorous but perhaps more ecologically important species now under threat. They also are worried that deextinction could provide a psychological safety net that might lead people to take the current extinction crisis less seriously. Why spend millions to save a species if it can be brought back later through the magic of genomics?

The counterargument offered by advocates of deextinction is that bringing some of these remarkable species back to life would supercharge the public's interest in the natural world. It would relieve some guilt and also might create some optimism. It would certainly be "exciting," to use one of Paul Crutzen's favorite terms for earth management in the synthetic epoch. Coming face to face with a woolly mammoth in an ecosystem from which they had been absent for five thousand years sounds like a prospect simply too cool to resist.

Each of these different sides of the deextinction coin has merit to them. The ethics of deextinction is complex. But apart from these questions about animal welfare, ecosystem balance, and conservation priorities, a deeper background issue characterizing the Synthetic Age emerges. Deextinction presents us with a dramatic choice about a direction. Like nanotechnology and synthetic biology, interventionist techniques like deextinction bring human design deeply into processes that once gave the natural world its shape. Evolution through natural selection and the extinctions that accompanied it are part of what made the world into what it is. These are some of the earth's most fundamental metabolic processes, which over countless generations shaped the planet on which we all were born.

Although humans have always nudged these processes in certain directions, either accidentally or deliberately, they never have so deliberately coopted the processes in order to remake the biological community this fundamentally. They never have made nature quite this synthetic. Reversing extinction is a radical form of artificing—turning the planet's composition of species into whatever we decide suits us best. It is not just a matter of shuffling around the species already here. Deextinction makes choices about which new—or old—species to place where. Ecosystems become increasingly artifactual, a product of our choices. Nature as inalienably "other" from us slowly starts to slip from view.

John Stuart Mill's reflections on nature in the nineteenth century were not only about whether humans were entirely inside or outside its realm. He also suggested that nature functions as an important grounding substrate for our lives. This substrate serves, he declared, as "the cradle of our thoughts and aspirations." As Mill understood it, the world into which we were born formed the essential and unchosen backdrop against which humanity over many centuries forged its identity and its sense of self.

In an essay published in *Orion Magazine*, environmental writer Scott Russell Sanders picks up on a similar theme:

> the warmth of sun on our skin, the stroke of wind, the sound of thunder and rain, the push of rivers and swell of seas, the smell of thawing dirt, the sight of leaves and blossoms unfurling, the pinpricks of light from stars, the intake of breath and thump of heart. These sensations have yielded humankind's perennial images for the ultimate nature of things, imagery that runs through scriptures, folktales, petroglyphs, poems paintings, and other symbolic expressions the world over.[11]

The "ultimate nature of things" has always been assumed to originate in something distinct from us. Its fundamental workings were dictated by larger geological, ecological, or divine forces. We had to accept it for what it was and were compelled to find ourselves a home within its inescapable embrace. For people like Sanders and Mill, this natural cradle offered a doorway into life's meaning.

Today's attempts to remake ecosystems mean new territory for both the earth and our species. If we take the path of assisted migration, gene drives, and deextinction, this planet is no longer simply the landscape and ecology into which we are born. It becomes the synthetic system we choose to construct. This would not just be true in close-to-home settings such as the city, our sub-urban living spaces, and our agricultural environments. It would also be true in the wild.

Gene drive pioneer Kevin Esvelt and his colleagues see exactly what is at stake. "The ability to manage ecosystems by altering wild populations," they say, "will have profound implications for our relationship to nature."[12] Humanity's designs, not nature's, would become our cosmic cradle. As a philosophical or religious prospect, there is something distinctly unnerving about this change. The replacement of nature with artifice can presumably go too far. Sanders is alert to the dangers. "The realm of artifacts," he says, "for all its ingenuity and convenience, becomes pathological if that is the only world we know."[13]

For increasing numbers of ecologists and land managers, this is simply the hand that our growing populations and escalating impacts have dealt us. We have no choice but to make the earth into a well-designed artifact. The separation between natural history and human history, as some commentators have put it, is over. Nature, in the mind of Jedediah Purdy, must be added to the list of things that are no longer natural. Our species now has to call all the shots.

But some fuzzy ethics are going on here. The suggestion that we have no choice represents a substantial leap in logic. Although it may be true that everything is now *influenced* by humans, it does not follow that every feature of the natural world must now be *determined* by humans. In fact, some of our influence remains relatively insignificant. In a number of places, nature still operates largely independently of human intent. Our species' role remains negligible. Such places are highly valued across numerous cultures and religions, and there are hundreds of millions of environmentalists who have been laboring to keep it so.

There also is a big difference between a premeditated and an inadvertent change. We may have inadvertently affected much of the world through pollution, the accidental introduction of nonnative species, and climate change, but we have not yet set about intentionally shaping the whole planet. The latter represents a whole new level of commitment. We have not yet made this choice, and it does not appear that we are compelled to do so. The idea that humanity might endeavor to make all global change premeditated is without precedent and sets an extremely high bar for the performance of any would-be planetary engineers.

When thinking about whether we can reach this bar, we ought perhaps to remember that there is an inherent wildness to the living world that always lies in wait. As Darwin pointed out, animal forms and behaviors are constantly changing over time. Although he did not understand the mechanisms, Darwin knew that the biological world had a propensity to shift continuously and in unpredictable ways. It will certainly retain this propensity, even in a Synthetic Age. Biological systems will always be subject to the random mutations that attend the phenomenon of descent. This residual unpredictability will ensure that would-be ecological engineers are in all probability going to receive some nasty surprises.

• • •

As provocative as all this talk of deextinction is, an equally tantalizing genomics discussion surrounds the sequencing the DNA of the Neanderthal, a close relative of modern humans that is thought to have died out about forty thousand years ago. A Swedish researcher named Svante Pääbo, impressed by the success of the Human Genome Project but more interested in extinct rather than living hominids, led a study at Germany's Max Planck Institute in the early 2000s to sequence the complete genome of the Neanderthal. The researchers published their results in 2010, offering an initial draft of the Neanderthal genome, and they followed that up with a more detailed one three years later.

Pääbo found that most of today's humans—with the exception of Africans—already have Neanderthal DNA making up between 1 and 4 percent of their genome. In the twenty thousand years or so that modern humans and Neanderthals shared the landscape in Europe and in Asia, a number of cross-species romances occurred. Some of these Neanderthal gene sequences have proved advantageous enough that they are still being selected for within modern humans.

Pääbo's current goal is simply to compare the genome of modern humans with that of the Neanderthal in order to understand what makes us different and what enabled us—rather than them—to end up as such a dominant species. A future step could involve using this comparative analysis and CRISPR editing techniques to reconstruct from a modern human's genome the full four billion base pair genome sequence of the Neanderthal. When this happens, some big ethical decisions will need to be made.

The somatic cell nuclear transfer techniques proposed to reverse the extinction of animals such as woolly mammoths and the Pyrenean ibex could in principle also be used for Neanderthals. If we decide to proceed, there is no question who the surrogate parents would be. Although deextincting the woolly mammoth would be exciting, deextincting the Neanderthal has an altogether more chilling feel.

We no doubt would hesitate before taking this step. Apart from all the philosophical questions it raises, there are considerable practical ones. How would the details be handled? Should the call go out to an adventurous woman to have Neanderthal DNA transplanted into one of her evacuated eggs? Or should two human embryos have altered germ cells injected so that they could become chimeras who might later combine to give birth to a Neanderthal? In either case, the resulting child would embark on the strangest imaginable life, orphaned from the rest of her species by about forty thousand years of intervening hominid history.

• • •

The need to make choices about *which* species should be saved *where* in a climate-stressed world is already here. For cold-adapted species like whitebark pine and bull trout, the writing may already be on the wall. We have changed too much at this point to imagine that an untouched natural order will remain intact and faithful to its historical state. Even without the changes that already have occurred, the certainty of future climate change demands from us painful decisions about which species we are going to invest in saving. It also demands choices about what sort of adaptive techniques are acceptable.

In some cases, saving species will be impossible without captive breeding and relocating struggling individuals. It also may involve constructing ecosystems that are more climate-resilient by carefully piecing together different combinations of living and nonliving elements to provide a buffer against changing conditions. Reintroducing beavers in appropriate areas, for example, can create better water storage in drying ecosystems during hot summer months. In a climate-stressed world, we might embark on new levels of ecological engineering in order to create the surroundings that will support conservation goals. Using gene drives to change the DNA of wild species might also have some conservation, as well as humanitarian, appeal. The use of genomic technologies to resurrect extinct species or even to design completely new ones is the next frontier we will have to decide whether to cross.

In each of these cases, the idea of nature as our sacred native inheritance is rejected. As with other technologies of the Synthetic Age, we do not settle for what we find around us. We rebuild nature as we see fit. To take up this role deliberately and on a global scale is completely new territory for our species. For a few more years, we have an opportunity to decide collectively which parts of this future we want and which parts we might reject. If there is no honest and open conversation about these matters, then the only interesting question remaining will be how far and how recklessly down that untrodden path we will travel.

7 The Evolutionary Power of Cities

Nanotechnology, synthetic biology, assisted migration, and deextinction all promise to impress human designs onto the natural order of things at ever deeper levels. These technologies provide some of the most startling illustrations of what a Synthetic Age might bring. They reach deeply into the planet's metabolism and substantially reconfigure its workings according to our designs. These rapidly growing powers to transform nature at the most fundamental levels put a new type of power in our hands, demanding what Australian science writer Gaia Vince has called "an extraordinary shift in perception."

Not all the significant frontiers being crossed today, however, involve dramatic new techniques. Other phenomena suggestive of a new period in the earth's history are more familiar and more gradual, caused not by a disquieting new technical development but by the accumulated consequences of trends that have been building for some time. These other types of transformation are in some cases no more than the inevitable product of our nature as hardworking and social hominids. They are much less technical and much more mundane. But even if they do not require new developments in the nanotech or molecular biology lab, they are no less transformative. A potent example of such a phenomenon is urbanization.

Sometime in 2007, a person was born in a city somewhere on the globe who tipped the proportion of *Homo sapiens* that lives in cities over the 50 percent mark. Despite the fact that cities cover only 2 to 3 percent of terrestrial surface area, more than half of humanity is now urban-dwelling. There is no going back. The human condition is now inevitably and increasingly that of a city dweller.

Our species evolved on the savannahs of Africa. For close to 200,000 years, we lived in grasslands and scrubby forests, hunting and foraging, using skins, wood, and grasses for shelter. We were subject to nature's caprice and continually on the move. As ice, water, and other barriers to our roaming tendencies came and went, we spread onto new continents, where we eventually started staying put and planting a few crops, gradually forming larger and larger associations for the advantages of mutual protection, efficiencies of labor, and perhaps also—one might hope—the joy of good company. For the vast majority of our history, we were a species that felt the dirt under our feet, experienced the changes of weather and season directly on our skins, and encountered face to face the animals and insects that shared our landscapes. This constant exposure to a living, breathing world shaped and selected us for a distinctive physiology, certain defining behaviors and dispositions, and a particular type of mind.

Since the birth of that urban child in 2007, *Homo sapiens* has become a species that has traded out the habitat in which it evolved. Urban living is progressively becoming the species norm. In 1800, only 2 percent of the human population lived in cities. By 1900, that portion grew to 15 percent. By 2050, the number will reach 80 percent. Humans increasingly occupy an evolutionarily unfamiliar niche, where the sensory and physical dimensions of a life lived in daily contact with the natural world have been replaced by a whole set of alternate experiences.

The significance of this change from rural to urban life should not be underestimated. Cement and traffic, 90 degree corners, bars, sirens, glass, and streetlights increasingly dominate our senses. Droves of people hustling by in cars and in buses, on foot and atop skateboards, create an unrelenting and often joyous urban cacophony. More than 500 billion metric tons of concrete now coat the surface of the earth, more than two pounds for every square meter of the planet's land and sea. The places we spend the bulk of our time are constructed by urban designers and corporate decision makers rather than by evolutionary forces. Joining us in this new normal are a sundry assortment of rats and raccoons, cockroaches and crows, foxes and other urban critters all opportunistically seeking out what they can exploit in this reconstructed concrete world. Although not selected for it by evolution, the city is where most of our species now dwells. *Homo sapiens* has become *Homo urbanus*.

Many aspects of this urbanization are highly desirable. Cities can offer significant freedom from the back-breaking work often associated with labor in the countryside. They present new opportunities for prosperity and virtually unlimited possibilities for companionship. They foster artistry and offer inspiration through the large numbers and types of personalities who often parade colorfully in front of us in the course of a typical urban day. Due to their density, cities also create efficiencies that are unavailable in the suburbs and the country. Urban living attracts because its anonymity can be a refuge for those who desire to escape from a troubled past. It also can provide second chances for those who fell on their face the first time around.

There is little doubt we are a highly adaptable—perhaps the most adaptable—species. This means that few of us consciously feel the force of this rupture from our evolutionary past. But despite the obvious allure, *Homo sapiens* in a city remains to some extent a primate out of place. As far as our genes are concerned, we live in an alien world. Phobias about snakes slithering out of toilet bowls, coyotes snatching children out of strollers, and diseases infiltrating city water supplies reveal the location of our biological roots. Endless coffee-shop discussions about record snowstorms or stifling heatwaves demonstrate the power that the idea of raw nature continues to hold over the urban mind. The support for environmental causes typically found in the urban areas of developed countries suggests a deeply harbored fondness for a disappearing past. The shadow of the wild continues to haunt the psyche of even the most entrenched urbanite.

Alongside the human transplants, fast-breeding and opportunistic species are changing their behaviors and their genomes so that they will fit better in the urban world. City-dwelling swallows are evolving shorter wings that allow them to avoid the traffic better. Moths are gaining different color patterns so that they have more suitable camouflage in their new concrete habitat. Evolutionary forces are turning city-bound mice into separate subspecies in different city parks, unable to exchange genes with cousins who live a few blocks away. Sparrows and starlings have raised the pitch of their calls to compensate for the background urban noise. The urban world is not just pushing *Homo sapiens* in different directions. It is doing the same to the rest of the biota. There may be no reason to lament this urban path given its many positive contributions to our humanity. But there is no doubt it is a path causing an unstoppable shift in who we, as well as the species that like to live alongside us, essentially are.

• • •

A second and related agent of evolutionary change is the progressive banishment of darkness from the world at the hands of electric light. Paul Bogard has written poignantly of his deep regret at the "end of night." He points out that the spread of electricity across many parts of the globe has condemned real darkness to the planet's history. This lack of night comes with sizable biological consequences. Excessive illumination is disrupting the natural rhythms created by millions of years of the earth's steady axial rotation.

The first photos of the earth from space taken by lunar-bound astronauts revealed a spectacular blue marble poised in front of a star-speckled expanse. The individuals lucky enough to see the planet from this vantage point were all transformed. Edgar Mitchell memorably described his impression of it as "a small pearl in a thick sea of black mystery." The planet's finitude, its swirling beauty, and its apparent fragility gave our species its first clear sense of our lack of astral significance. Norman Cousins later remarked that "what was most significant about the lunar voyage was not that man set foot on the Moon but that they set eye on the Earth."

More recent photographs of the earth taken at night have revealed a pearl that is increasingly crossed by spider webs of yellow light projected from cities and the transportation corridors between them. The world is now comprehensively illuminated. Thanks to the ubiquity of electric light, less and less of the planet falls genuinely into darkness any more. Power shunted through incandescent filaments, the gases of fluorescent lights, and a billion light-emitting diodes means that darkness is being pushed off the landscape by this electric interloper. Synthetic light races through the air for miles beyond its intended destination, leading to a diffusion rate that far exceeds that attainable by the bulldozers and diggers that make its spread possible.

Prior to Thomas Edison's design of the first commercially viable light bulb, nighttime illumination came only from flames fueled by imperfect sources, such as wood, whale oil, paraffin, and natural gas. The light from these sources danced unpredictably and was always mottled by the smoke of imperfect combustion. The spread of the light was limited by available fuel, environmental conditions, and a basic lack of penetration. Many still feel attached to the light provided by a cavorting flame, seeking it out in wood and wax when wishing to disappear into memories or create venues for intimacy.

When the limited light cast by these flames was overtaken by that of incandescent bulbs, the nighttime started to change its color from a deep inky black to various shades of orange, yellow, and white. The carefree spreading of megawatts of unused light into the night sky has led to a pale dome of illumination above every population center. This glow refuses to leave the city's vault even when most of its residents are asleep. Bogard quotes an Iroquois writer who told him "we have the night so the Earth can rest." As electrification has spread across the world, the amount of rest available to the earth has diminished. This loss to the planet also appears increasingly to be a loss of our own.

Human bodies have natural circadian rhythms. These rhythms are adjustments to the waxing and waning of light during the earth's daily rotations. Evolution lodged such patterns deeply inside of us. The circadian rhythm has an influence on hormone production, body temperature regulation, blood pressure, and other key functions. Plants, animals, cyanobacteria, and fungi all have similar rhythms that are their own evolutionary adaptations to the rising and setting of the sun. Leaves turn to face the sun and drop in the fall, petals open and close daily, animals rest, and bacteria fix nitrogen at rates that are direct responses to periodic and predictable changes in light. When patterns of light and darkness change, organisms must rapidly adapt or pay the price.

Consider that more than a fifth of all mammal species are bats. In addition to these well-known lovers of a dark world, 60 percent of invertebrates and 30 percent of vertebrates are nocturnal. This means that a large number of the living forms that share the planet with us have evolved so that darkness is an essential factor in their well-being. Of those species that are not fully nocturnal, a large number are crepuscular, a word that has exactly the right sound to describe the creeping and partially hidden character of activity that takes place at twilight.

The swapping out of darkness for light across much of the planet affects all of these species. Sea turtles emerging from the surf and no longer able to navigate by the moon due to beachfront floodlights are perhaps the best-known victims of artificial illumination. But in addition to the turtles, millions of other species are shifting their behavioral patterns to accommodate a planet that is increasingly lit up.

Peregrine falcons, for example, are adapting to the new frontier of urban living by figuring out how to hunt pigeons, ducks, and bats in the city at night. The nocturnal hunt no longer involves the 200-mile-per-hour "stoop" from above that has made peregrines famous as the fastest birds on earth. Illumination provided by the glowing city means that the nighttime ambushes involve a new type of stalk. Peregrines fly upward toward the illuminated bellies of their unsuspecting prey, rotating at the last second to pierce the hapless victim's feathered breast with their deadly talons. Like *Homo sapiens* adapting to the city, peregrines are figuring out ways to live, feed, and rest in a world that no longer resembles the one their genes prepared them to find.

Bogard is concerned about how little research exists on the human health consequences of the disruption of circadian rhythms. In developed countries, up to 20 percent of the workforce is employed in service industries that require employees to be awake for large portions of the night. Night-shift workers such as janitors, health care attendants, and those who labor in twenty-four-hour manufacturing facilities are some of the people who bear this burden. Those who work the graveyard shift seldom replace the number of hours of sleep they missed at night with the same number of hours of sleep during the day.

In a striking indication that the end of night has consequences, the World Health Organization's International Agency for Research on Cancer concluded in 2007 that "shift-work that involves circadian disruption is probably carcinogenic to humans."[1] It is thought that this may have something to do with disruption to the production of the hormone melatonin, but at the moment, this is little more than a guess. It should come as no surprise that the human body has a deep biological connection to the earth's diurnal rhythm.

One of a growing number of local and national organizations concerned about the loss of darkness in America is the National Park Service. This agency has created a "night sky team" to raise awareness of the importance of darkness as a new type of resource, pointing out with impeccable logic and federally approved rhythm that "half the park happens after dark." In 2006, the Park Service committed itself to preserve the natural lightscapes of parks, which it described in ethical language as "resources and values that exist in the absence of human-caused light." Artificial light is now deemed an "intrusion" into the park ecosystem, suggesting that the distinction between what is artificial and what is natural is not yet completely moot.

Astronomers too are obviously miffed. Light pollution from cities is making optimal conditions for star gazing harder and harder to find. This is not only the concern of a few professionals with big budgets. Astronomy may be one of the most widely enjoyed arts on the face of the planet, ranging in its practitioners from PhD scientists with multi-million dollar telescopes to five-year-old children trying not to topple to the ground while craning their necks upward to wonder at the night sky. Seeing the moon and the stars above is one of the most orienting of human experiences. It was recently determined that more than a third of the world's population can no longer see the Milky Way due to the presence of light pollution. "If we never see the Milky Way," asks Bogard (quoting Bill Fox), "... how will we know our place in the universe?"

Urbanization and the spread of artificial light are transforming life for all of earth's species, not just our own. These transformations are taking place incrementally—one more baby born in Surat, one more streetlight erected in the Sudan. But the aggregation of all these tiny impacts creates a fundamental new reality for biological beings. The fact these transformations are the gradual result of millions of separate decisions made by individuals not connected to each other by language or ideology makes their consequences no less significant.

In addition to growing cities and glowing night skies, other subtle shifts are in the wind. The very air surrounding us now bumps with electromagnetic waves carrying the information that sparks cell phone calls, internet searches, and evenings spent with streaming media. The apparent stillness of the air is increasingly an illusion. The intangible medium that fills a raptor's wings and surrounds our skin hums with the energy of millions of anthropogenic memos being processed by the nearly billion transistors manufactured each day.

Nor are the oceans skipped over in this energetic refurbishing of the earth. As marine waters acidify due to their absorption of carbon dioxide from the atmosphere, the ability of the oceans to mute low-frequency sounds decreases. This means that noise travels farther underwater as a direct consequence of the burning of fossil fuels. As a result, the oceans are increasingly percolated by a growing mélange of sounds. Marine life such as dolphins and whales that are highly dependent on acoustic signals are finding their most basic communications more and more disrupted by these changes in resonance. At the same time, melting sea ice also lets more sunlight into the upper layers of the ocean, creating a world that is both lighter and noisier for those swimming through northern waters.

The planetary transformations also reach high. The thin band of space that lies just above our atmosphere is now sliced and diced by tons of metal and silica from functioning and defunct satellites. More than 500,000 pieces of space junk are currently tracked by the

National Aeronautics and Space Administration, many of them hurtling by at more than 17,000 miles per hour. Astronauts interested in photographing the blue marble from space today have to dodge these orbiting projectiles. They must plan carefully in advance with their earth-bound minders in order to avoid a catastrophic collision with some of this debris.

In each of these cases, the world has little by little been transformed into something different. It has become increasingly shot through with the consequences of our practices and our technologies. The idea that there would always be an endless blue sky, an inky black night, a silent ocean, or the infinite emptiness of space is increasingly a distant memory. The transformations are deep enough that even natural selection itself is fast becoming unnatural.

These incremental and inexorable changes caused by small decisions spread over countless locations across the globe are in some ways more insidious than those caused by dramatic new technologies. Their causes are distributed thinly among billions of people, and the resulting changes are almost always unintentional. Nobody planned to make the oceans noisier by burning fossil fuels or to crowd the air with electronic signals. Nobody launching a weather satellite wanted inner space to become a swirling blender filled with speeding metal parts. Like many impacts to the planet that result from the pursuit of better lives, these changes have crept up on us through no particular ill will on anyone's part. Yet the world shaped inadvertently by humans is the world in which we now wake up. It is the illuminated, wired, and increasingly urban world that we and the species around us must learn to live within.

In her reflections on this new period in the earth's history, Gaia Vince sounds resigned to the new normal. Longing for the Holocene past, she says, is a waste of time. One of the primary lessons she thinks we must learn is that we no longer have the luxury of refusing our new role or of rejecting our new surroundings. We instead must make our choices about how to shape our surroundings more deliberately and dispassionately. She admits, however, that we are left with few tools to guide us in these decisions. This deluge of planetary-scale changes has created wholly unfamiliar challenges for the scientific, cultural, and religious philosophies that used to "guide our place in the world." Despite our being cut so philosophically adrift, Vince still thinks that the nature of our time leaves us no option but to boldly assume the role of "masters of our planet."

The alternative visions of "masters of the planet" or "plain members and citizens of it" are the perennial haunts of environmental thinking. A strong preference for the latter drove Aldo Leopold's quest for a land ethic three-quarters of a century ago. There is no doubt that in Leopold's time the context was different and the tools at hand were not as sharp. Less had already been transformed, and significantly less technological potential lay at our fingertips. We were not yet capable of reaching as deeply into the planet's workings.

Despite the advance of time, Vince's talk about the inevitability of becoming masters and the need for dispassion is far from satisfactory. We *could* choose to experience the regret that Vince does not allow herself to feel. We *could* resolve to do things differently if we so chose. Unlike the earth-shaping cyanobacteria that first created an inhabitable atmosphere by making oxygen available in the Archaean and Proterozoic eons, humans have the capacity to look around and contemplate the changes we are causing. We can ask how deeply we want to manipulate the planet and make a thoughtful decision about how synthetic we want our surroundings to be. We can weigh how much we value the idea that some parts of nature must remain independent of us.

The biggest tragedy would be for us to allow these decisions to be made on our behalf, without our input, by market-driven forces. It is a terrible misunderstanding of technological change to assert that markets simply reflect what people want. Markets move down the channels that promise the most reward to those who figure out how to exploit them. Decisions about those channels are not made with the public's needs or interests in mind. They are opportunistic decisions based on financial promise.

In *Walden*, Henry David Thoreau declares that we are rich in proportion to the number of things we decide to leave alone. Bill McKibben invoked Thoreau a century and a half later when he implored us to say about certain forms of technology "Enough!" If our species fully abandons the idea of leaving certain parts of the natural world alone, letting evolution be *natural* evolution, and letting the sky and the land in some places remain quiet and still, then the game of preservation that has motivated environmental thinking since Thoreau is over. With the ending of that game, the idea that some part of our association with the natural world should be one of deferral and restraint is abandoned. An orienting ethical hinge will have been abruptly detached from its anchors.

After this happens, there will be little limit to what we will feel entitled to change. Broken free from any constraints, there will be no reason to hold back. At that point, our gaze will likely shift upward to include not just the pliable and ever shifting environments that surround us by land but also the skies that envelop us from above. Tinkering with the atmosphere will start to look like the next logical step in the rapidly expanding realm of the Plastocene.

8 How to Turn Back the Sun

From the moment former vice president Al Gore stood on a stage with his laser pointer, his goofy graphs, and his somewhat wooden delivery for the documentary *An Inconvenient Truth* (2006), there has been a widespread recognition that climate change presents humanity with a massive economic and moral migraine. When humans alter the climate under which they live, something literally earth-shaping is going on. Every inch of the world is transformed. No longer simply "the domain of the gods," the skies above us become something we have made. And when climate changes, everything changes.

Recognition of the seriousness of climate change was slow to arrive. Throughout most of the Holocene, carbon dioxide comprised only 0.028 percent of the atmosphere. All the carbon dioxide in the atmosphere together would make a layer only 3 millimeters thick if spread equally across the surface of the planet. It did not seem that any increase in the concentration of a gas that formed such a small part of the gaseous whole could make much of a difference. Warnings from atmospheric scientists fell on deaf ears. For more than a quarter century, for reasons that amounted almost entirely to self-interest, a number of developed nations doggedly refused to admit that anything was awry.

In recent years, even in the most foot-dragging parts of the world, most leaders have acknowledged that redirecting natural history by slow-cooking the planet with atmospheric carbon levels unseen for at least three million years—and perhaps as many as fifteen or twenty-five million—is probably not a good idea. The ones that have not acknowledged this are widely recognized to be living in a fantasy world increasingly detached from reality. After watching an ever-growing newsreel full of super-typhoons, record floods, and crop-scorching summers, the global community is finally realizing that nobody wants a planet with only one white ice cap. Yet that is certainly the world we are heading for as Arctic sea ice goes on a rapid downward spiral. In my home state of Montana, we are trying to work out what to call a protected area currently named Glacier National Park that within a couple of decades will lack any year-round ice. Figuring out how to minimize the effects of global warming presents humanity with one of history's most formidable social and economic challenges, challenges that recent United Nations meetings in Paris, Marrakech, and Bonn, have attempted to address.

Capturing the scale of these challenges, a rather clued-in academic named Stephen Gardiner has called climate change the "perfect moral storm." Gardiner is a midcareer philosopher teaching at the University of Washington in Seattle. British by birth and educated at Oxford and Cornell, Gardiner is one of those people who carries around with him the air of being completely in control of the conceptual terrain. This is largely because he is.

A serious man who can look a bit like George Clooney when he grows a short beard, Gardiner is not averse to showing he means business by wearing blue blazers and cufflinks at conferences where the radicals are wearing ratty T-shirts and Crocs. He has held visiting fellowships at Oxford, Princeton, and the University of Melbourne. Even with all this raw, native ability, Gardiner remains a popular person in his field, engaging students and faculty colleagues with intensity, while retaining both his manners and good nature.

For a decade or more, Gardiner has been a leading figure in the area of research known as climate ethics. This means he is an expert on the rights and wrongs of what got humans into the climate mess in the first place and on the various strategies being offered for how to get us out. His idea of climate as the "perfect moral storm" conjures up the story of the ill-fated fishing boat *Andrea Gail* in the Hollywood film *The Perfect Storm* to create the image of several unusual forces that converge to make climate change difficult to solve. The three big storms Gardiner identifies are the global storm, the intergenerational storm, and the theoretical storm.

First, the global storm. Gardiner points out that the *global* dimensions of climate change are psychologically difficult to grasp because at a gut level it seems so improbable that simply driving a car to the store or turning up the heat by a few degrees can have real consequences for people living on the other side of the world. We are not used to thinking that our everyday actions have global effects. Next, he suggests that the *intergenerational* elements of the storm confound us because we can barely imagine what life will be like next week, let alone for our descendants fifty or two hundred years from now. Finally, Gardiner notes the presence of a *theoretical* storm, which amounts to a total lack of adequate political and ethical theories that might help us get out from under such complicated and ominous threats.

Gardiner's bottom line is that the global problem that created anthropogenic greenhouse gases is unprecedented and we have very few good ideas about how to deal with it. His prognosis is grim. In that movie about the fishing boat, Gardiner's rough-shaven doppelganger, George Clooney, was unable to avoid a watery grave. This does not bode well. With glaciers and ice caps melting, the perfect moral storm of climate change is in the process of bestowing a similar fate on millions. "Good luck, chaps!" you can imagine Gardiner uttering with a polite smile and a friendly wave as he retreats to his faculty office to continue working on these ideas.

Hunkered down in the face of this perfect storm, those charged with doing something about climate change realize they have to become considerably more creative. Back in the 1960s, an American nuclear physicist named Alvin Weinberg coined the idea of a "technical fix" to describe the deployment of an engineering solution to take care of the most intractable social problems.[1] If you cannot create change personally and politically, then maybe you can do it technically. A good illustration of what Weinberg meant by a technical fix was the invention of antilock brakes for cars. Because people show a persistent inability to drive slower on wet and icy roads, then at least their numbskull driving can be made slightly less dangerous by equipping cars with antilock brakes. With this technology, putting the brake pedal to the floor when driving too fast *may* not lead to the outcome the driver deserves. Antilock brakes, then, are a technical fix that to some degree can mitigate the difficult social challenge of making people drive slower. Science, in other words, comes (partially) to the rescue.[2]

If climate change really is as morally and politically challenging as Gardiner suggests, is it also the type of problem that might allow for a technical fix? Perhaps a scientific genius can dream up an engineering solution that will permit us to continue our carbon-intensive ways. We live in an era of unprecedented technical manipulation of the surrounding world. It is conceivable that a technical solution to climate change will succeed where politicians and activists have so far failed.

It should be obvious that the threats posed by climate change will require considerable scientific and technological advancement if the challenges are to be met. As Bill McKibben likes to point out, burning dirty black rocks and oils dug out of the ground at considerable risk to miners and rig workers may have been a master stroke of genius in Roman times (and even for a few centuries beyond that), but it does not seem like a technology fit for the age of self-driving cars and Instagram. Cheaper and more efficient solar panels, wind turbines, and alternative transportation methods are all technological innovations that will be necessary to make a serious dent in carbon emissions. More effective batteries for energy storage, breakthroughs in building and urban design, and the development of smarter power grids are also part of the package of technical innovations that will be required if the world is to successfully transition to clean energy.

Technologies such as these, however, may not be the only options on the table. Some contemporary climate thinkers have started to discuss the possibility of another sort of approach to climate change. According to this line of thinking, some of the fundamental workings of the atmosphere are tweaked in order to bring down temperatures. This is the vision of climate engineering.[3]

Before he ever said a word about climate engineering, Paul Crutzen had already left his mark on the history of atmospheric science with his warnings about how the chemicals used in refrigeration and other industrial applications were depleting the ozone layer. The 1987 Montreal Protocol, an agreement negotiated by the United Nations on the back of some of Crutzen's work, had arrived just in time to save the earth from a major planetary disaster. The 1995 Nobel Prize he shared for this work with Mario Molina and F. Sherwood Rowland made him for a time the world's most cited author in the geosciences. Combine this with Crutzen's earlier findings about how the use of atomic weapons could precipitate a devastating form of artificial winter, and you get a person who is as much of a household name as any atmospheric chemist can probably ever claim to be.

Crutzen turns out to be the Steve Jobs of the atmosphere, simply unable to stop leaving his mark in that sphere. After drawing the world's attention to the problem of the disappearing ozone layer, Crutzen began another important discussion in 2006, suggesting that the global community should start to take seriously the idea of using technology to cool the climate artificially. In the absence of good prospects for reducing greenhouse gas emissions any time soon through political means, Crutzen advised in a 2006 article for the suitably named journal *Climatic Change*, research into "artificially enhancing Earth's albedo" should begin in earnest.[4]

Albedo is a measure of how reflective something is (as opposed to being a measure of how white a rabbit is, although whiteness, it turns out, can be involved in both). It can be thought of as a shininess factor. If you increase something's albedo, then a larger portion of the energy that strikes it is reflected back in the direction from which it came. If you could somehow do this on a scale proportionate to the whole earth, then a meaningful chunk of the solar energy that currently enters the atmosphere and gets trapped there could be intercepted and sent back out into space before it has the chance to heat up anything. The result would be a slight reduction in global temperatures.

Shortly after Crutzen went public with this proposal, a bevy of prospective climate engineers gained the confidence to emerge out of the shadows. It is one of the hazards of winning a Nobel Prize. People start listening to you. The discussion of climate engineering almost immediately took off at a high rate of knots.

In order to increase the earth's albedo, a range of possible solar radiation management (SRM) methods can be employed. Among the most futuristic (and expensive) is the idea of putting millions of small mirrors into orbit in order to reflect back sunlight before it reaches the upper atmosphere. In addition to the expense and the technological complexity of this idea, no single country has the sort of space program that could be devoted to such a project at this point in time. As a result, the idea of orbiting reflectors was almost immediately put on the back burner.

Another much less technical suggestion is to paint large portions of the earth's surface white while at the same time using genetic modification to lighten the tint of common crops to make them more reflective. Albedo modification at ground level is certainly simpler and less costly than trying to do it in space. But since two-thirds of the earth's surface is ocean, unless you were also willing to lighten the oceans—perhaps by infusing them with trillions of microbubbles or covering them with a layer Styrofoam peanuts—these whitening strategies would likely be fairly ineffective.

The SRM proposal that has drawn the most attention so far is the idea of putting some form of reflective particle or droplet into the stratosphere to intercept solar energy before it gets any closer to the earth than a very high-flying jet. The result would be a type of hazy atmospheric barrier to incoming sunlight that would measurably increase the planet's albedo. This scheme is nothing if not bold. Like modern-day versions of Cnut, the eleventh-century king of Denmark who placed his throne defiantly on the beach and ordered the tides to retreat, climate engineers would be executing strategies that attempted to do no less than turn back the sun.

The massive advantage that the stratospheric particle strategy has over all the other approaches to SRM is that, without question, scientists know it would be effective. What makes them so sure? It is not because people have ever tried to deliberately cool the planet in the past. All of humanity's ill-fated tinkering with the atmosphere so far has been completely accidental. Scientists know it would work because they have tracked how big volcanic eruptions throughout history have caused exactly the same effect.

When the Indonesian volcano Krakatoa blew in 1883, it created what was reputed to be the loudest noise in history, a noise that was heard more than three thousand miles away. Barograph records suggest that the sound wave caused by the eruption circled the globe three and a half times. The eruption obliterated the island on which Krakatoa was located and caused a tsunami that killed over 36,000 people. Debris ejected from the crater was thrown up to fifty kilometers in the air. Skeletons of the volcano's victims floated across the Indian Ocean on pieces of pumice and, more than a year later, washed up on the shores of east Africa, accompanied by numerous fat and happy crabs.

A large portion of the rock and the dust thrown into the air came back down to earth almost immediately. In fact, a number of temporary islands were formed in the ocean around where the original Krakatoa had stood. However, a portion of the dust and gases propelled upward in the eruption made their way into stratospheric air currents, where high-altitude winds dispersed them around the globe. Here they lingered for several years. Sunsets half a world away glowed with fantastical colors as the trillions of particles ejected during the eruption refracted the sun's rays. Norwegian artist Edvard Munch is said to have painted his famous work *The Scream* after witnessing the intense evening sunsets caused by Krakatoa as far away as Christiania harbor in Norway. But much more climatically significant than the temporary enhancement of sunrises and sunsets was the fact that the explosion noticeably cooled the earth.

The sulfur dioxide gas spewed from the volcano resulted in the forming of trillions of droplets of sulphuric acid in the stratosphere. These droplets temporarily increased the albedo of that normally transparent layer of air and reflected some percentage of the incoming sunlight directly back out into space. Reliable historical records indicate that Northern Hemisphere temperatures fell by at least 1.2 degrees Celsius in the summer following the eruption. Snow fell in unusual places, frost impacted spring plantings, and summer harvests were either delayed or did not happen at all. In addition to the lower temperatures, record rainfall struck the west coast of the United States, and precipitation patterns across the globe were disrupted. Worldwide temperatures did not return to normal for several years, until the suspended volcanic debris had all precipitated back down to earth as a mild acid rain.

The case of Krakatoa is far from unique. After Tambora blew in 1815, the world experienced "the year without a summer," a year nicknamed "Eighteen Hundred and Froze-to-Death." Mary Shelley is said to have been inspired to write *Frankenstein* in part because she spent most of the summer of 1816 confined indoors by the lousy weather in a villa on the shores of Switzerland's Lake Geneva. She and the friends she was cooped up with—including Lord Byron, a man who at the time was reputed to be "mad, bad, and dangerous to know"—reportedly scared each other half to death by inventing horror stories to pass the time as wind and sleet rattled the tiles above their heads.

Mount Katmai in 1912 in Alaska, El Chichón in Mexico in 1982, and Mount Pinatubo in the Philippines in 1991: all of these large volcanic eruptions resulted in measurably cooler temperatures being recorded around the world. A small cottage industry has arisen among climate scientists trying to match historical temperature dips to the timing of large volcanic explosions. From mini-ice ages to massive global extinction events, the connection between the projection of large quantities of gases, liquids, and particulates into the sky and the lowering of worldwide temperatures is widely recognized. Volcanoes, in other words, are the "proof of concept" for solar radiation management. They are Paul Crutzen's "Exhibit A."

The question now being asked is whether, in our desperation to do something about climate change, humanity should try to create the same effect artificially.

• • •

David Keith is a likable, extremely bright Canadian engineer who currently divides his time between the University of Calgary and Harvard.

Twenty years of research has earned him a reputation as one of the foremost advocates of technical fixes for climate change. Keith is a wiry man with a somewhat impish grin and boundless intellectual energy. In conversation about his work, Keith comes across as the kid who has just discovered the best toy in the store. In addition to being a first-rate physicist and engineer, Keith also has a philosophical side to him as well as a deep passion for wild and frozen places. He doesn't just want to know how to engineer and build things, he also wants to know why we should engineer them and what tradeoffs they will demand from society. He thinks hard about how technologies can transform fundamental components of the human relationship to nature.

As part of his attempt to wrestle with these philosophical puzzles, Keith went to Missoula, Montana, after graduate school to read complicated texts on Martin Heidegger with renowned philosopher of technology Albert Borgmann. For months, Keith picked Borgmann's brain to understand how building things shapes society in subtle and unnoticed ways. Montana was the perfect setting for him to think through these puzzles. Although he was training to be an engineer, much of Keith's life was oriented around a deep love of the wild. On crisp fall weekends, he roamed the backcountry with a shaggy-looking bear biologist named Chuck Jonkel to learn what he could about the habits and the habitat of the region's famous bruins. When winter snows blanketed the landscape, he strapped on a pair of skis and headed deep into the backcountry.

Since being named one of *Time* magazine's "heroes of the environment" in 2009, Keith has taken on a set of rapidly escalating responsibilities. These include being executive chairman of Carbon Engineering, which seeks to pull carbon directly out of the atmosphere using giant fans and sorbent chemicals; mentoring a team of Harvard post-docs in technology and public policy studies; and helping to distribute money from Bill Gates's Fund for Innovative Climate and Energy Research.

When thoughts of the frozen north become too enticing, Keith has been known to shut down his lab and disappear on a ski trip across remote Arctic landscapes for three weeks with a couple of his friends.

On those long days traversing the crusty surface, he contemplates how climate change means not only the loss of ice and snow for the Arctic but also something even harder to recreate—the loss of nature's pure wildness. Back in his laboratory, he is trying to do something about it.

Keith is one of the world's leading experts on engineering stratospheric aerosols. When his book *A Case for Climate Engineering* came out in 2013, Keith was invited by *The Colbert Report* to present the idea of managing solar radiation through mimicking volcanoes to a national audience. Unfortunately for Keith, the reflective agents most suitable to put in the stratosphere to shield incoming sunlight are droplets of sulphuric acid. Colbert, of course, excoriated him.

Keith: You put, say, twenty thousand tons of sulfuric acid into the stratosphere every year, and each year you have to put a little more in. And this doesn't in the long run mean that you can forget about cutting emissions. We will need to reign in emissions.

Colbert: (*sarcastically*) No, we'll get to it eventually. In the meantime, we're shrouding the earth in SULFURIC ACID.

Keith: So people are terrified about talking about this because they're scared that it will prevent us cutting emissions.

Colbert: (*suspiciously*) Right. ... And also that it is SULFURIC ACID.

Keith: It is.

Colbert: Is there any possible way this could come back to bite us in the ass? Blanketing the earth in sulfuric acid? Because I'm all for it. This is the all-chocolate dinner. I still get to have my CO_2, and I just have to spray sulfuric acid, right? All over the earth.

Keith: Right question, but we put fifty million tons of sulfuric acid in the air now as pollution. It kills a million people a year worldwide.

Colbert: (*feigning stupidity*) That's good or bad?

Keith: It's terrible.

Colbert: But it will be better if we put more in?

Keith: We're talking about 1 percent of that. A tiny fraction of that.

Colbert: But if it kills a million people ...

Keith: It's bad.

Colbert: We only do 1 percent more, we're just killing ten thousand more people.

Keith: You can do math, okay. But that's— So killing people is not the objective here.

Colbert: Killing people is not the objective. I just wanted to be clear.

Colbert gloated. Every attempt Keith made to present solar radiation management as a serious response to climate change was met with derision. Colbert found it too easy to present Keith as a caricature of a mad scientist for the sake of entertaining the crowd.

Keith's book, however, poses significant questions. If you take seriously the amount of harm that unchecked global warming will cause, if you recognize that these harms will fall disproportionately on the global poor, if you acknowledge that these populations are not only the least equipped economically to cope with climate change but are also the least implicated in the rise of greenhouse gases in the first place, and if you concede the undeniable reality that conventional strategies for reducing the harms of climate change are not being implemented quickly enough, then there seems to be a strong moral case for doing something dramatic. Managing solar radiation by faking volcanic eruptions, Keith insists, could be the only way to ensure justice for those on track to suffer the most through climate change.

After Paul Crutzen's high-profile endorsement of climate engineering in 2006 broke the taboo against discussing the subject, the debate about this potential response to warming temperatures quickly turned into an orgy of point and counterpoint from ethicists, government experts, legal scholars, atmospheric scientists, and ecologists. The prospect was too mouth-watering, and too terrifying, to ignore. Suddenly, everybody had a view about the pros and cons of the outlandish prospect of deliberately engineering the climate.

Climate engineering through stratospheric aerosols has all the hallmarks of a classic technical fix. It displays a riveting techno-geek prowess. It contains the prospect of Crutzian atmospheric-knight-in-shining-armor heroics. And it allows everyone to heave a gargantuan sigh of relief at the prospect of solving the world's most vexing problem (while simultaneously preventing Al Gore from ever needing to pace a stage with his laser pointer again). It would also, by all accounts, be surprisingly cheap and technically simple to do. Climate engineering looks like a signature technique for a Synthetic Age.

So why haven't we started already?

• • •

As Colbert cheekily pointed out, the technology comes packaged with some serious concerns. Keith himself describes stratospheric aerosols as "cheap," "fast," and "uncertain." There is no doubt that they could be deployed relatively cheaply compared to the cost of decarbonizing the whole global economy. Keith offers a ballpark estimate of about $1 billion a year. There also is no doubt that the reduction in temperatures could be relatively swift, perhaps within a matter of days or weeks. Unfortunately, as Keith readily admits, the full effects of using solar radiation management on the climate are not completely known.

Global climate is a delicate balancing act. Stopping sunlight (short-wave radiation) from entering the atmosphere at the top does not completely compensate for preventing accumulated heat (long-wave radiation) from exiting it through the growing layer of greenhouse gases. Any time you change the *radiative balance* of the atmosphere—the relationship between heat coming in and heat going out—you are altering a range of phenomena, including the rates of evaporation of water from the oceans, wind patterns, temperature gradients between places, and plant productivity. Reflecting sunlight on a global scale is a major shake-up of a highly complex and unpredictable system. It creates lots of uncertainties.

Rainfall is a particularly worrying concern. The connection between the explosion of Krakatoa in August 1883 and the following summer's deluges in California suggests that moisture patterns are significantly disrupted by tweaks to the albedo of the stratosphere. With so many of the world's poor located in either arid or flood-prone regions like sub-Saharan Africa and Bangladesh, any changes in precipitation could come with a high human cost. Existing patterns of rainfall have been integrated into the rhythm of people's lives. Many of the world's small-scale subsistence farmers depend on regular seasonal monsoons for their crops. Even though research suggests that, on balance, the effects of solar radiation management could be positive, the uncertainty about what it would do to rainfall creates a dark cloud over Keith's moral argument. Climate would likely remain wild and unpredictable, even when engineered by the world's best experts.

More precise scientific knowledge of the effects of solar radiation management is clearly desirable, but this knowledge currently depends almost entirely on the predictive accuracy of computer models. Unfortunately, the global effects of SRM technology cannot be tested convincingly without actually being deployed. Actual deployment would be a rather high-stakes test of a new technology designed to do something as dramatic as changing the climate for the whole planet. There remain serious doubts about whether scientists could ever, even in principle, know enough to manage sunlight. The *butterfly effect* suggests that tiny perturbations in one part of the system can create dramatic unforeseen disruptions in another. Trying to manipulate a complex, chaotic system on a global scale sounds like it could be a fool's game.

To make matters worse, in an environmentally conscious age, the idea of spraying chemicals into the atmosphere from planes absolutely terrifies people. A British proposal for a small-scale test of a mechanism that would have sprayed nothing but water from a balloon anchored 1 kilometer off the ground was abandoned in 2012 due to a range of public concerns. Although these concerns about spraying less than two bathtubs of water were probably an overreaction, a real test of solar radiation management could be a legitimate cause for worry. If particles were put into the stratosphere as a test, high altitude winds would rapidly spread them around the globe. If something unexpected happened, it would be impossible to collect them again. The research necessary for determining the safety of stratospheric aerosol deployment is therefore not only incomplete but also highly controversial. Keith is painfully aware of these issues as he and a colleague at Harvard, Frank Keutsch, try to get their own test of the effects of putting a frozen water mist into the stratosphere from a high-altitude balloon—followed by a later test with calcium carbonate particles—off the ground.

The worries about solar radiation management are not limited to questions about storms, droughts, and rainfall. SRM is also unable to deal with what is sometimes called "the other CO_2 problem"—ocean acidification. As atmospheric concentrations of carbon rise, the oceans absorb more and more of this colorless, odorless gas.[5] The carbon dioxide absorbed at the surface of the ocean reacts with the water to form carbonic acid, which then disperses throughout the marine environment.

Carbonic acid has already turned the oceans significantly more acidic, with the impacts greatest where the water is coldest. Marine ecosystems have seen an increase in acidity averaging around 30 percent in the last two hundred years. Some Arctic areas will see a more than doubling of acidity by the end of the twenty-first century. The impacts on marine life are profound.

Carbonic acid makes it much more difficult for marine organisms to grow shells. One of the most widely known effects of ocean acidification is that coral reefs all over the world are in a death spiral, bleached by rising temperatures and inhibited by the inability of the organisms that compose them to form their skeletal structure. As a consequence, the fish that use the reefs are losing their spawning habitat, and the complex food webs on which the ocean ecosystem depends are in disarray. Other ecologically important species sensitive to ocean chemistry—including plankton, seaweeds, and oysters—are also threatened by carbonic acid. Continued ocean acidification will transform the oceans to a state barely recognizable to today's marine scientists.

A final worry arises with the realization that while engineers are tinkering with the chemistry of the stratosphere, the earth's natural volcanic processes are continuing at their own pace. Even if synthetic particles are sprayed by reluctant engineers high into the stratosphere, tectonic plates at the surface of the earth will continue to grind against each other, and red-hot magma will continue to rise. Should an eruption on the scale of a Krakatoa or a Tambora occur at a time when the stratosphere is loaded up with aerosols, the earth would suddenly get a double dose of cooling. These depressed temperatures would last for several years and cause a devastating sequence of crop failures. The nuclear winter that Crutzen researched in the 1980s could find itself with an accidental competitor for our species' annihilation.

All of this means that the magic bullet of solar radiation management could hit a big piece of volcanic pumice and ricochet off in an unknown direction. Although SRM could helpfully reduce average global temperatures, it comes with the concussive one-two punch of continued ocean acidification and a range of uncertain climatic effects. Both of these are high prices to pay for what was designed to be a simple technical fix to climate change.

David Keith is a smart man and knows all this. So did Paul Crutzen when he first broached the subject in 2006. But in the absence of other good alternatives, both of them still think this is a bargain that might be worth striking. The most straightforward option for dealing with climate change—a slow and relatively painless transition to low-carbon sources of energy—is already off the table, thanks to the procrastination and distrust that have plagued the last three decades of international climate policy. The angles for dealing with climate change are increasingly narrowing. In this perfect moral storm, not only is the weather becoming threatening, but the options are getting progressively fewer.

The scientific questions raised about solar radiation management are significant, but an additional challenge for the prospect of mimicking volcanoes is the pancreas-exploding politics that is sure to attend any plan to begin active solar radiation management. Who could be trusted to develop such a potent global technology? How much SRM should the global community choose? Whose fingers would be allowed to touch the global thermostat? These governance challenges are substantial, and it is not clear how the voices of those with a stake in future temperatures—in other words, all seven and a half billion of us on the planet—would be heard.

The potential for climate engineering to cause massive geopolitical instability is not lost on its advocates. Will those countries that appear to have something to gain from warmer temperatures, like Canada and

Russia, be happy to revert to their frigid climates of the past? Would regional partners or even individual nations try to deploy climate engineering in their own interests without prior discussion with other parties? Where would China stand? Would the interests of the poorer countries be ignored? It is hard to envision an internationally just procedure that could acceptably answer all these questions. New think tanks have been formed whose sole purpose is to wade through this political morass.

In response to these difficult politics, Keith's project at Harvard is committed to being multi-disciplinary and inclusive of different viewpoints. It will involve governance organizations, environmental NGOs, and various organizations from civil society. The idea is to introduce skeptical voices and ensure maximum transparency. From the outset, the project is designed to help figure out what the pieces of a politically acceptable, international effort at solar radiation management might look like. Despite these types of assurances, some critics have quite plausibly suggested that SRM is either an inherently undemocratic technology or an inherently ungovernable one or both.

• • •

Alongside the pressing scientific and governance questions that are raised by the prospect of turning back the sun, a number of more abstract puzzles are giving the growing number of philosophers interested in this problem some major heartburn. These puzzles bring us face to face with the conceptual shock to our sense of self and surroundings that an unrestrained Plastocene epoch threatens to deliver.

One of the most defining of these puzzles is the question of what sort of entity an earth with an intentionally manipulated climate would become. The past ten thousand years have offered a relatively stable background climatic context for everything—both good and bad—that has happened to humanity. All of our central events—the domestication of animals, the invention of agriculture and writing, the birth of the world's major religions, the building of the pyramids and the Great Wall of China, the Renaissance, the Enlightenment, the two world wars—have all taken place under a generally even Holocene climate. As John Stuart Mill might have put it, this climate has been civilization's reliable cradle. But now, 150 years later and living on a warming planet, this familiar background is shifting.

Bill McKibben has drawn attention to this more clearly than anyone. In 1989, at the start of public awareness about a climate crisis, McKibben wrote a book with the arresting title of *The End of Nature*. In just over two hundred fact-filled and reflective pages, McKibben poignantly lamented that greenhouse gases emitted from the burning of fossil fuels threatened to turn the whole of the earth into "a product of our habits, our economies, our ways of life." Thanks to climate change, the planet is utterly different now. We have turned every spot on the earth into "something man-made and artificial."

McKibben suggests that we now live on "Earth 2.0," a new planet enveloped by a new atmosphere. An earth surrounded by dangerously elevated levels of greenhouse gases becomes, at some profound level, a different place. "We have built a greenhouse," says McKibben "where once there bloomed a sweet and wild garden." Everything that remains is to some extent human-influenced.[6] What McKibben keyed into from the start was that this change represents a massive loss to our species. The Holocene climate represented "the separate and wild province, the world apart from man to which he adapted, under whose rules he was born and died." These rules were now changing.

But if, as McKibben supposes, we already live on Earth 2.0 thanks to inadvertent climate change, what additional psychological payload does climate engineering saddle to our backs? Some think that climate engineering would effectively turn the earth into a *giant artifact,* a planet from here onward intentionally managed by humans to reflect and absorb exactly the right amount of solar energy. This would mark a whole new period of history in which humanity deliberately takes control of the planet's geophysics.

In this new epoch, the fundamental properties of the earth's relationship to the sun would cease being as fundamental as they have been in the past. Patterns known as Milankovitch cycles, which have varied predictably over the 2.5 million year Pleistocene epoch that preceded the Holocene, have in the past determined the cycles of heat and ice that shaped the earth's ecology. In a climate-engineered world, these very slight extensions and compressions of the planet's elliptical orbit, the changing angle from the vertical on which its axis spins, and the varying directions in which that axis points would no longer be the determinants of global temperature. These small but influential planetary wobbles would become largely irrelevant. Unlike any other planet we know, the earth's inhabitants themselves would moderate the amount of solar irradiance their surroundings received. Atmospheric engineers would turn dials and employ algorithms to ensure that only a mathematically determined amount of heat would be allowed to warm the earth's surface at any one moment. For us, the solar system would become a *solar-calibrated system*, whose thermodynamic properties would be continually tweaked in order to make our lives more comfortable.

Like a tireless potter eternally shaping her clay, humans would become responsible for perpetually molding climate. This would be more than just shaping a particular ecosystem or landscape on a local scale, something that has been done throughout human history. With climate engineering, people would assume continuous management of everything under the sun. They would take control of one of the most basic of planetary processes, one that previously was determined by autonomous forces originating deep within the physics of the solar system.

Assuming this role would increase human responsibility considerably. Most ethicists and legal theorists would assert that doing something intentionally is far different from doing something accidentally. Think of the difference in blame between murder and manslaughter or the difference between throwing a rock at someone intentionally and accidentally loosening one during a hike on a steep slope.

As far as the responsibility it creates, climate engineering is like throwing the rock. It involves the deliberate intention to change the atmosphere, not the careless polluting of it. It opens up an entirely new chapter for the home planet and a new kind of global responsibility.

Going beyond Earth 2.0, which had an unintentionally altered climate, a climate-engineered planet would be Earth 3.0, or perhaps it would become something we could no longer called earth at all.

Some advocates for the technology have acknowledged that the phrase *solar radiation management* conjures up images from science fiction movies and conveys a false sense of control. They have attempted to rebrand the SRM acronym as *sunlight reflection methods* to suggest that climate engineering is somehow more benign than the idea of global-scale solar radiation management. It is clever wordplay but probably futile. McKibben, for his part, wants to remain focused on emissions reduction. He has described the whole climate engineering discussion as "annoying" and the psychological impulse behind it as "dubious." The idea provided by this technical fix is unlikely to go away any time soon, however. At this point, the giant earth-artificing cat seems to be out of the rapidly warming bag.

In many ways, the management of solar radiation is the quintessential activity of the Plastocene, with humans reaching into the heart of one of the ultimate natural processes and making it theirs. What could be more indicative of a Synthetic Age than a species intentionally shaping its own planet from the atmosphere on downward? Forget terraforming Mars. We can do it here first.

Paul Crutzen, the theorist who started this discussion, saw from the beginning how intimately linked climate engineering was to the prospect of a new epoch of human history. He recognized the challenges but also saw the potential. We no longer occupy a time when humans can stand back and expect the earth to manage itself. Today's challenges demand "appropriate human behavior at all scales," he suggested, "and may well involve internationally accepted, large-scale geoengineering projects, for instance to 'optimize' climate."[7]

Can clever technicians reconstruct the world by optimizing climate? It sounds like a plot from a science fiction movie. But it is not, and it is worth paying attention. With climate harms escalating by the week and the political will to do something about it far from guaranteed, a decision about whether to head down the climate engineering path may lie not far in the future. With leaders of some powerful countries distinctly prone to seduction by the sublime beauty of their technologies, it may take nothing short of an extraordinary mobilization of concerned citizens to persuade eager governments to turn away from this path.

9 Remixing the Atmosphere

The rather alarming prospect of spraying acid into the stratosphere has garnered most of the public attention that has been directed toward climate engineering. But another solar radiation management technique is attracting interest from researchers. This one involves increasing the brightness of marine clouds to reduce the amount of heat absorbed by the ocean.

Clouds appearing within the first few thousand feet or so of the ocean surface can be enhanced by spraying a mist of salt water through specially designed nozzles from fleets of slow-moving boats. These brighter clouds would reflect sunlight back toward the upper atmosphere. Practiced on a large enough scale, this cloud-enhancement technique could create an effect that is similar to the deployment of stratospheric aerosols by reflecting back a meaningful percentage of incoming solar energy before it can warm the dark ocean beneath it.

Although the practice of cloud brightening is still in the modeling stage, advocates of this technique note how satellite photos from NASA show ship exhausts promoting the growth of clouds that stretch for hundreds of miles across the ocean. The engineering technology required to do this with salt water rather than diesel emissions is thought to be relatively simple. The toughest challenge is getting the sprayed particles to be consistently the optimal size. After this is mastered, fleets of self-piloting boats could be deployed to track grids across portions of the ocean surface while spewing a salty mist into the sky.

The idea of marine cloud brightening does not push people's buttons as much as the prospect of spraying aerosols into the stratosphere does. People seem to be a little less intimidated by the idea of tinkering with puffy white clouds at sea than they do by the idea of shooting acid droplets into the stratosphere. The ocean also has a more familiar and perhaps more comforting connection to human history than the band of stratospheric air ten miles above us does. When the future of the planet is at stake, a few more clouds over the ocean does not seem like such a high price to pay.

Some of the immediate safety worries of climate engineering are also considerably reduced with marine cloud brightening. Cloud brightening is much more easily stopped than high-altitude aerosol deployment. If the spray nozzles are turned off, the ocean clouds disappear in a number of hours or days. One of the lessons of Krakatoa, by comparison, is that an injection of stratospheric aerosols continues its effects for several years. Marine cloud brightening also poses no threat to the earth's protective ozone layer, but stratospheric aerosols potentially do. Stephen Colbert would probably be one of the first to point out that ocean cloud brightening involves spraying salt water into the air rather than sulfuric acid. This too provides some reassurance.

Yet practiced at a large enough scale, ocean cloud brightening is still a form of global solar radiation management. Like stratospheric aerosol injection, it too messes with planetary albedo on a scale that could cause problems. The two big worries attending stratospheric aerosols—uncertain precipitation impacts and continuing ocean acidification—haunt even this less glamorous form of climate manipulation. Public resistance to marine cloud brightening, although not currently as great as the resistance to stratospheric aerosols, may in the end fall not too far short of the resistance to injecting chemicals into the stratosphere.[1]

Both stratospheric aerosols and marine cloud brightening are focused on the idea of albedo modification. Albedo modification has drawn a lot of attention, but it is not the only option on the table for managing the atmosphere. Climate engineers have other tricks up their sleeve. These alternatives approach the problem from a different angle. Instead of attacking solar energy, they attack carbon.

At virtually the beginning the climate engineering discussion, an influential report by Britain's Royal Society divided the field into two main types of technology.[2] The first was solar radiation management. The second was a range of techniques that would suck carbon dioxide out of the atmosphere and store it somewhere safe for the long term. This strategy is known as *carbon dioxide removal* (CDR).

Because the Royal Society's labeling decision gave this latter technique the same "geoengineering" tag as the more glamorous practice of solar radiation management, carbon dioxide removal has thus far been forced to play second fiddle in the climate engineering orchestra to its attention-grabbing cousin. Since the climate agreement in Paris in 2015, however, CDR has attracted increasing amounts of attention. The Paris agreement made it clear that if humanity is to stand any chance of keeping global temperature increases from preindustrial times below 2 degrees Celsius, not only have we got to stop putting carbon into the atmosphere, but we also have to start taking it out.

There are many ways to pull carbon out of the atmosphere. One is simply to plant more trees. This low-tech solution does not sound very radical and is in fact unlikely to solve the climate problem on its own. It requires an awful lot of land, an awful lot of trees, and a way to ensure that the carbon released from dying or harvested trees did not make its way straight back into the atmosphere. Concerns about the land grab that this massive tree planting would demand mean that although trees tend to be a welcome part of the discussion of carbon storage, this strategy is not being treated as a stand-alone solution to climate warming.

Another biologically driven way to draw down carbon dioxide is to generate massive blooms of phytoplankton in the oceans. This can be done by spreading powdered forms of vital elements such as iron, potassium, or phosphorous on the ocean surface in areas that are otherwise nutrient deficient. With these additional ingredients introduced into the soup, phytoplankton naturally occurring at the ocean surface will proliferate and take up increasing amounts of carbon dioxide as they photosynthesize.

As the primary producers in the oceans, large numbers of these phytoplankton will quickly enter the food chain. As a result, some portion of the carbon taken up by these microorganisms eventually will end up sinking toward the deep ocean either in the form of feces deposited by the trillions of marine organisms that have consumed the phytoplankton or in those same organisms' bodies when they die. The incessant snow of carbon, it is hoped, will end up in long-term storage in sediments on the ocean floor.

The phenomenon of ocean fertilization with nitrogenous materials is something that was already built into marine nutrient cycles for much of evolutionary history. In the days before the world's whaling fleets depleted the numbers of these giant, free-roaming cetaceans, whales ably performed this fertilization function with their feces. The nutrient-laden bowel movements of the ocean's largest inhabitants is thought to have had a measurable influence on the global climate by stimulating the growth of carbon-sucking microorganisms.[3] With tens of millions of these marine mammals no longer happily defecating in the upper levels of the marine ecosystem, the distribution of ocean nutrients today falls far short of what it was in the past. This creates another good reason for protecting and enhancing whale populations, one that could be added to the numerous motivations for safeguarding these complex and charismatic kindred species.

One of the problems with artificial ocean fertilization, however, is that the jury is still in recess over just how much carbon such microorganisms can actually absorb when nutrients are sprinkled on the ocean surface. There are also doubts about whether the carbon really ends up in safe, long-term storage on the ocean floor. Other concerns center on the broader ecological effects of spreading of nutrients throughout the marine ecosystem. Like climate itself, marine food chains are complex things, and this type of chemical intervention would likely create significant side-effects that could spiral off in unanticipated directions.

Some of these effects could be subtle. Climate activist Naomi Klein, after hearing about an illicit ocean fertilization experiment that had taken place near her home on the British Columbia coast, wondered whether the unusual sightings of orcas nearby signaled an ecosystem already tipped into disarray by spreading the equivalent of Miracle-Gro on the ocean surface. Echoing McKibben's remarks about the atmosphere, Klein lamented the possibility that after these sorts of interventions "all natural events can begin to take on an unnatural tinge."[4]

In working as a fisherman, I have experienced the otherworldliness of watching dozens of humpback whales feeding around our vessel as we passed through Frederick Sound near Petersburg, Alaska. There was no direction to look without seeing flippers and flukes slicing the smooth surface and clouds of spray put up by blowholes and the whales' repeated breaching. The thrill it created was due in part to the sense of the wildness and spontaneity in the spectacle. The large-scale dumping of nutrients across the ocean surface might in future make such experiences feel more like a semi-orchestrated SeaWorld display than a window into nature at its most unconstrained.

Because many of the biological techniques for sucking carbon on a large scale come with ecological risks, alternative carbon dioxide removal methods are being proposed that involve scrubbing carbon directly out of the atmosphere on land using chemical rather than biological means. One of these methods involves artificially enhancing the process by which rocks naturally weatherize.

The weathering process of mountains by rainfall is one of the primary mechanisms of the carbon cycle, and it has been responsible for drawing massive amounts of carbon out of the atmosphere throughout the earth's history. Rainfall is always mildly acidic. As a result, it creates a slow but significant reaction when it falls on rocks. The mildly acidic rain causes silicate and bicarbonate ions to run off from terrestrial surfaces into streams and rivers. Some of this runoff percolates down into subterranean caves and fissures. Above or below ground, these ions become grand masters at grabbing carbon.

In subterranean settings, they precipitate out of the water to create carbon-rich stalagmites and stalactites, whose pointed features continually rake the cool breezes that move through underground caves. The ions in the surface waters end up in the oceans, where certain marine organisms use the carbonate ions to create the calcium carbonate that comprises their shells. Microscopic algae named diatoms use the silicates for constructing their cell walls. At the end of their lives or the lives of their consumers, the bodies of these organisms rain down to the ocean floor, where as compacted sediments they slowly turn into dolomite, limestone, and other rock types. Locked into these lithic impoundments, they serve as a long-term vault for about six trillion tons of atmospheric carbon. It is thanks to these natural weathering processes that the earth had an atmosphere capable of supporting life in the first place.

Enhanced rock weatherization can sound like a bizarre strategy for dealing with climate change, but it also has a certain logic to it. If climate change is an artificial acceleration of one part of the natural carbon cycle through digging up and burning fossilized carbon fuels, then speeding up the process that puts carbon back into the ground looks like it could be a smart response. The chemistry of enhancing the weatherization process is not complicated. Spreading a common natural mineral called olivine over areas of rock will speed up the rate at which the silicate and carbonate runoff occurs. As a result, huge amounts of additional carbon will be drawn out of the atmosphere.

If running a large-scale chemical process on the bare surface of exposed mountainsides and mesas gives you pause, then a different carbon-sucking process could be located on artificial structures that reach sixty feet up into the sky. The *direct air capture* (DAC) of carbon would involve using an engineered structure that is a modern day combination of a windmill and a ship's sail. These structures, euphemistically known as "artificial trees," would capture carbon from the breeze as real trees currently do when they photosynthesize. Such synthetic trees would need to be distributed widely across the landscape in places where they were constantly exposed to the ambient air. As this ambient air passed across them, a chemical reaction on the surface of their "leaves" would gobble atmospheric carbon. The carbon harvest then could be extracted from the chemical before being transported and stored somewhere safe, perhaps in the geological formations from which oil and gas had been extracted.

Direct air capture has the most clinically engineered feel of all of the carbon dioxide removal strategies. To solve the intractable carbon problem, you apparently just need to precision-engineer the right device. By creating an artificial version of nature's best-known carbon-scrubbing organism at a large enough scale, it might become possible to draw sufficient carbon out of the atmosphere to start making a real difference. Nature's trees could accept a helping hand from the deployment of more efficient artificial ones. The hopes expressed at the Paris Conference that we might learn to be good at pulling carbon out of the atmosphere might start to be realized. David Keith, not satisfied with just the one geoengineering role testing solar radiation management technologies at Harvard, is also involved in a company called Carbon Engineering that is working to develop and commercialize an effective version of DAC.

• • •

There is a lot to like about carbon dioxide removal as a climate strategy. First, it seems to address the root cause of the greenhouse gas problem in a way that solar radiation management does not. Although SRM masks one of the major effects of carbon dioxide by reducing temperatures, it does not tackle the primary cause of those elevated temperatures, which are the greenhouse gases themselves. Carbon dioxide removal, on the other hand, goes directly for the cause by removing a climate-warming gas from the atmosphere.

Carbon dioxide removal has another happy consequence. As a result of the attention to root causes rather than symptoms, CDR would slowly start reducing the dangers of ocean acidification in a way that solar radiation management cannot. Less carbon in the atmosphere means less carbonic acid in the world's oceans. Coral reefs would begin to repair themselves with untold benefits for the nine million or so species that depend on them. Crabs and oysters would get to keep their shells.

The warm and fuzzy feeling associated with carbon dioxide removal continues. If carbon dioxide is considered to be a pollutant, then CDR is simply a type of pollution capture and removal. What's not to like about that? We all have a responsibility to clean up our own messes, whether that mess appears on the ground or wafts overhead in the mixed-up layers of the sky.

In contrast to most technical fixes, carbon dioxide removal also appears to have a reassuring naturalness to it. Oceans, forests, algae, phytoplankton, and rocks all take carbon out of the atmosphere, as do many types of bacteria. Humans might think of themselves as honoring their biological roots if they started doing the same thing on a grand scale. Bacteria and spruce trees do it. Perhaps we should be following their lead.

There is one final advantage to carbon dioxide removal that is becoming increasingly attractive by the day. CDR has the important advantage of not only being able to reduce the effects of the carbon now being emitted into the atmosphere but also starting to remove the carbon that has been emitted in the industrial past. There are already in excess of 400 parts per million of carbon dioxide in the atmosphere, up from 280 before the start of the industrial revolution and only 330 as recently as 1975. The rate at which *Homo faber* has been putting this damaging gas into the sky has doubled since the 1970s. Most climate scientists and most of the politicians and diplomats at recent climate summits agree that this is already far too much. One well-known climate organization calls for a maximum atmospheric concentration of carbon dioxide of 350 parts per million. The earth has already overshot this number by a substantial margin.

Carbon already in the atmosphere is the gift that keeps on giving with many of its effects lasting for thousands of years. If we are to restore the atmosphere to acceptable concentrations of carbon, it will be necessary not only to reduce current carbon emissions but also to go after the carbon that has already been emitted. Carbon dioxide removal can do this. Solar radiation management cannot. SRM leaves all emitted carbon in the atmosphere until it is naturally reabsorbed, a process that takes many thousands of years. This delay is certain to ensure great hardship and suffering for many of the world's vulnerable people as well as for numerous at-risk species.

These fundamental differences between carbon dioxide removal and solar radiation management are striking. If CDR really is a cousin to SRM (as the Royal Society's labeling suggests), it is only a very distant cousin. In the age of earth-shaping technologies that can sometimes look ominous, CDR appears to be a refreshingly wholesome endeavor—far less of a desperate stop-gap measure than SRM and far more of good planetary hygiene.

• • •

But not all the news about carbon dioxide removal is comforting. Perhaps the most disconcerting piece of news is that it is not clear that the required level of carbon removal is technologically or economically feasible on anything like the timeline required. None of the various strategies under consideration have yet been demonstrated to be viable at the appropriate scale even though the majority of pathways under consideration for keeping rising temperatures at a manageable level already rely on them.[5]

Direct capture of carbon from the air would also place huge demands on land and other natural resources, requiring the development of a whole new industrial-scale infrastructure similar in size and ecological impact to today's oil and gas industries. There would be colossal manufacturing and transportation requirements as well as massive demands for energy and fresh water if the various chemical scrubbing processes were to work effectively.

There also would be a high aesthetic cost. Artificial trees would lack the beauty of real trees and would provide little in the way of habitat for birds and insects. The carbon-scrubbing machines that Keith's Carbon Engineering firm proposes look like a cross between the worst type of 1960s office building and a giant hovercraft. Stacked metal modules containing giant fans would move ambient air across surfaces drenched in the carbon-capturing liquid. The carbon scrubber would need a nearby power source to keep the fans turning, and it would be surrounded by pipes and other infrastructure to move the saturated liquid somewhere for processing.

The visual impact of a carbon-scrubbing future could be significant. In addition to seeing rows and rows of power-generating wind turbines spread across the landscape, we also would see forests of carbon-scrubbing towers reaping their carbon harvest. The machinery required to keep the sorbent surfaces exposed and to move carbon-saturated liquid away would create a constant noisy whine. Fields of such structures would clearly be a fine technical achievement, but they also would be an aesthetic nightmare. The service roads, infrastructure, and ecological disturbance they would create seems likely to make today's fights over the siting of wind turbines seem like child's play. Only the knowledge that the infrastructure was designed to solve the problem of atmospheric carbon (rather than create it) would provide any sort of aesthetic compensation.

Many types of carbon dioxide removal also would be expensive and potentially disruptive to existing economies. Artificial trees would not be cheap to manufacture. Large-scale afforestation would displace significant amounts of food crops. Enhanced weathering of rocks would require olivine to be mined and spread across the landscape in enormous quantities. Some of the strategies might also have questionable legality. Ocean fertilization, for example, is already banned by the London Convention against dumping scary things at sea.

Carbon dioxide removal, then, is no panacea. Although the general idea looks like a move in the right direction and may be necessary in some form in order to reduce the concentration of the carbon we already have shoveled into the atmosphere, the various strategies being discussed face numerous technical and social barriers to implementation. One authoritative survey article of a range of CDR technologies dampens enthusiasm with a studied understatement: "It is far from certain that positive net environmental and societal benefits from CDR at very large scale will be achievable."[6] None of this bodes well for the idea that CDR might be the magic bullet that solves the climate problem.

Unfortunately, international climate policy is already placing a heavy reliance on carbon dioxide removal to make the broad targets agreed on in Paris achievable. In the Intergovernmental Panel on Climate Change's (IPCC) fifth assessment report released in 2014, a combination of biofuel production with attendant CDR technology known as bioenergy with carbon capture and storage (BECCS) was deemed absolutely necessary if the global community is to stand any chance of meeting its climate goals.

In the climate context, BECCS means switching from burning fuels that are fossilized to burning fuels that are grown. Because the carbon emitted by burning a fuel crop roughly equals the amount of carbon the crop has absorbed from the atmosphere during its lifetime through photosynthesis, bioenergy can theoretically come close to being carbon neutral.

A carbon-neutral energy supply is a good start, but it is hoped that new technologies will help us do even better. If the emissions created by burning the biofuels could be captured and permanently stored underground, then the process goes from being carbon *neutral* to being carbon *negative*, and the earth benefits from a net drawdown of atmospheric carbon. Bioenergy with successful carbon capture and storage achieves the highly desirable goal of *negative emissions*. It sequesters more carbon than it emits and reduces rather than worsens greenhouse gas concentrations in the atmosphere. Many of the national strategies that form the backbone of the international climate agreement forged in Paris make the assumption that BECCS will become widely used for energy supplies over the coming decades.

At this point, there are numerous obstacles to the successful implementation of BECCS. Questions about the appropriate crops, the energy and land-use requirements for their production, the politics of such a massive agricultural transition, and the right industrial processes for converting all that biomass into fuel all remain unresolved. Biofuels will certainly need to be much less carbon-intensive in their production than they currently are. The resulting products also will need to have adequate energy intensity. Airplanes cannot fly on mashed-up straw. The technologies required for capturing carbon from a biomass-burning power plant are not yet economical enough to implement. Although progress is being made, in its enthusiasm for BECCS the IPCC might be accused of counting its climate-saving chickens well before they have hatched.

• • •

A synthetic era in which humanity has awarded itself free reign to solve all of its major problems through technology offers some bold and tantalizing ideas for tackling climate change. In a full-throttle Plastocene, the climatic system as a whole would be open for manipulation through solar radiation management and carbon dioxide removal technologies. Many of these technologies are certainly worth exploring, particularly on the carbon dioxide removal side of the ledger. Airline magnate Richard Branson has created a Virgin Earth Challenge that promises a $25 million prize to the first organization to develop a safe, proven, and economically sustainable method for removing carbon dioxide from the atmosphere at an appropriate scale.[7] Researchers are champing at the bit for the opportunity to find that elusive technical fix for one of modern society's greatest challenges. Apart from anything else, huge amounts of money could be made if such a technology were commercialized.

Despite the enthusiasm in some quarters, there is a massive amount of uneasiness in others. Until now, the sky has seemed off-limits to intentional human interference in a way that the land has not.[8] There is no part of the earth's system that has remained more independent of *intentional* human interference than the atmosphere. With both carbon dioxide removal and solar radiation management, the idea of leaving the planet's thin atmospheric skin in a "natural" or "untouched" state is completely rejected. As climate activists from Al Gore to Bill McKibben have pointed out, by changing the climate we change absolutely everything. Setting out to deliberately manipulate climate would take humans into an entirely new realm. The synthesizing tendencies of the Plastocene would become planetary in scale. Traditional environmental thinking about the value of things independent of us would have been struck another blow. From the stratosphere on downward, we would be shaping it all.

Whatever reaction the prospect of climate engineering creates in a person—bone-tingling excitement or abject horror—the thought of it means looking at the sky in a new way. Not unlike the changes in perception that took place when astronauts first entered space, if climate engineering becomes the norm in the Plastocene, our association with the heavens would experience an irreversible shift. No longer would the sky be simply a distant starry firmament or an endless encircling vault. It would become just one more part of a managed system that humans would consciously and continually tweak to ensure our well-being.

Oliver Morton, author of a book on climate engineering titled *The Planet Remade*, understands the significance of this new role. Tinkering with the atmosphere, he says, "changes what it is for humans to be humans and what it is for nature to be nature—it takes human empire over the border of blasphemy." In a philosophically significant sense, the sky would be brought crashing down to earth as it entered the human orbit.

It is too late for philosophical or religious musings, climate engineering advocates insist. Our actions long ago compromised the integrity and independence of the atmosphere. Our only hope for forcing things back into shape is to go further down this path and reverse-engineer the atmosphere.

This response is not without its appeal. We have made a considerable mess of the atmosphere. Shouldn't we try whatever we can to clean it up?

But the extreme sort of planetary management that climate engineering involves comes with a huge price that has not gone unnoticed. Environmental writer Jason Mark has suggested that the deployment of climate engineering would create in us a type of "existential anxiety." On every day, at every hour, humanity would have assumed responsibility for whatever the climate was doing. Mark suspects this responsibility would cause us to tremble in fear constantly, worried that we might let "our grip slip from the string that keeps the planet in a semblance of balance."[9] Life in an epoch of juggling the climate would be a life lived perpetually on edge.

In a similar vein, Andy Revkin, a *New York Times* journalist who has written regularly on the climate crisis, has suggested that the prospect of climate engineering creates in us a "queasy mix of excitement and unease." It promises intoxicating power but comes with staggering responsibility. First there is the surge of adrenaline and then the vomiting.

At some moments, it is hard to judge if climate engineering is a Plastocene thinker's best dream or worst nightmare. It is probably a mixture of both. If humans could carefully and intentionally use an advanced technology to undo a giant unintended consequence of a certain lifestyle, then our species could celebrate its release from the grip of the climate crisis and issue a long global exhalation of relief. At the same time, humans could pat themselves on the back for possessing an ingenuity unmatched by any other species.

In this transitional moment, as our species gazes on the prospect of the Plastocene, we face real choices about how deeply into the business of rearranging the natural order we should go. The earth is struggling. We need to do something. Just how much to intervene is excruciatingly difficult to decide. Big choices about the expression of our powers are upon us. Whether to seize the reins and start proactively engineering the climate is one of the biggest choices of them all.

It is an uncomfortable position in which to find ourselves. Life was simpler when we still clung to the guiding idea of a pristine natural world that needed to be protected and left alone. Where previously we had to ensure only that our inventions and devices worked within culture, we now face the prospect of having to ensure that the biological, ecosystemic, and atmospheric processes work throughout the planet as a whole. Morton, who generally is supportive of climate engineering, identifies this as the inevitable price of turning what was once the province of the divine into a human responsibility.

There is no doubt that this represents a dramatic change. We would be taking on a lot more than we ever have before. The increasingly thin line that separates the human world from the natural world would have finally disappeared. Human history and natural history would begin to merge.[10]

And the more of these monumental earth system management tasks we take on, the more we will irrevocably change who we are.

10 Synthetic Humanity

The Synthetic Age will be distinct from all other epochs of our planet's history by virtue of the fact that human designs and desires will determine many of the earth's most basic functions. At times, these designs will be loosely guided by the way earth used to do things during the Holocene and previous epochs. At other times, humanity will strike out on a different path, determined to remake the world in a way that improves on what nature had provided. "It will be a marvelous challenge to see if we can outdesign evolution," said George Whitesides when talking about the promise of synthetic biology.[1] Attempting to "outdesign nature" from the atom all the way up to the atmosphere will result in an ever more unfamiliar planet.

In a full-throttle Plastocene, our species would play a role unimaginable by previous generations. Deep technologies promise the ability to recalibrate basic planetary characteristics, including the nature of matter, the arrangement of DNA, the composition of ecosystems, and the amount of solar radiation reaching the earth from the sun. Our descendants would be born into a world that the generation ahead of them had deliberately chosen to construct rather than one bestowed on them by geologic time. This would mark a change in the relationship between person and planet of a remarkable kind. There would be a tectonic shift in who we are and what we do.

Some eager futurists embracing these incredible new powers might wonder why the prospect of adopting this role should be any cause for hesitation. For many, remaking the earth seems an entirely reasonable next step in the story of *Homo sapiens*. If we have the knowledge and ability, why would we not take on an increasingly large role in shaping our surroundings? After all, that is what every other animal does. No species accepts the world as it is and leaves it untouched. Manipulation of the environment is a necessary feature of life itself. With their satellites and advanced computer models, humans are a special type of overseer. If we can work out how to manage the earth thoughtfully and skillfully, we might secure a better future both for ourselves and for the creatures that share the planet with us.

Such an attitude can seem realistic, informed, and practical. Worrying about a change in relationship between person and planet, by contrast, sounds just a bit too abstract and philosophical. This is not literally a change in us, only a change in what we choose to do. Adopting this new role does not mean growing two heads or sprouting wings. Despite taking on these new roles as the synthesizers of many of the earth's functions, the enthusiast for the Plastocene might say, we would remain fully human. We would simply be going further down the same path we had always been walking.

As we move further into the Synthetic Age, however, all such reassuring bets are off. Even a simple claim about remaining fully human may not always be true. The deep technologies we are developing for the surrounding world may, before long, be turned on ourselves. When this happens, confidence that our basic human essence will remain the same starts to run increasingly thin.

• • •

In May 2016, a closed-door meeting took place at Harvard Medical School that banned participants from tweeting the content of the discussions and talking with the news media. In attendance were 150 scientists whose purpose was to discuss the possibility of embarking on a genome project unlike any other that had taken place before. Organizers claimed that the secrecy was necessary because a peer-reviewed paper on the topic of the meeting was awaiting publication with a prestigious journal. That may have been true. On the other hand, the secrecy may have been due to the unsettling and provocative nature of the topic under consideration.

In the months leading up to the meeting, the techniques for synthesizing genomes from their constituent chemicals had become increasingly widespread in molecular biology. Up to this point, the technology had been used to synthesize only very simple genomes such as those belonging to bacteria and yeasts. But as the techniques for gene synthesis improved, it was becoming possible to think about constructing the longer genomes of increasingly sophisticated organisms. The meeting at Harvard Medical School was about taking the first few steps in the most complicated genome synthesis project yet—the synthesis of the entire genome of a human being. Participants were plotting how humans might build themselves from scratch in the laboratory, gene by gene, within a decade.

The techniques in synthetic biology that made this meeting possible were the same gene synthesizing and editing techniques that Craig Venter, Jay Keasling, and Svante Pääbo had already been using in their work on bacteria, yeast, and the Neanderthal. At the time of the meeting, state-of-the-art techniques were far from being able to stitch together the three billion base pairs of the human genome. The longest genomes so far synthesized were about half a million base pairs. The twelve million base pairs of yeast—the first organism with a cell nucleus and chromosomes to be a candidate for genome synthesis—was still a future dream. The human genome is 250 times as long as yeast and nearly six thousand times as long as anything in the bacterial world that had been synthesized before.

The organizers knew that the goal, at this stage, was entirely aspirational. Even if the technologies required for assembling three billion base pairs had already been available, it is doubtful anyone would have thought it ethically permissible to insert a synthesized human genome into the evacuated egg of a surrogate human mother. The defects that accompanied the birth of the Pyrenean ibex scientists tried to bring back from extinction in 2003 suggested that it would be unconscionably cruel to try inserting a full human genome into an enucleated egg. Despite these ethical doubts, the goal was judged to be valuable enough to warrant the gathering and to start plotting the various steps that might lie on the path ahead.

When various media outlets started to report on the meeting, the public response was largely one of disgust. Statements made after the meeting tried to distance the organizers from the ambition of actually creating a human being whose genomic inheritance came entirely from bottles of chemicals rather than the sex cells of living, breathing humans. They suggested that the main purpose of the project—a project they initially called the Human Genome Project 2—was simply to get better at gene synthesis.[2] They proposed that studying the technology could enable the development of future cells that were resistant to viruses or even the creation of protocells that might grow into human organs suitable for transplantation. Speakers at the conference also suggested that the synthesizing of complex cancer genotypes would allow for better models of disease, enabling more targeted genetic therapies.

Nor, they claimed while still on the defensive, was the project uniquely focused on humans. It would involve the synthesis of genomes for other animals, too, with the initial intention of creating functional cells rather than actual embryos. Mastering the challenge of putting genomes together would provide a range of benefits for humans as well as have considerable inherent scientific interest. The overall tenor of the organizers' damage-control efforts was consistent with Richard Feynman's famous remark several decades previously. In order to understand something fully, the first thing one had to do was to know how to build it.

Despite these efforts to tamp down the alarm, the immediate backlash both in the media and from other synthetic biologists suggested that the idea of synthesizing the human genome crossed some sort of unacceptable moral threshold. Drew Endy, a Stanford biologist who helped start the Biobricks Foundation and who is typically an enthusiastic supporter of synthetic biology, encouraged those involved to pause. "They're talking about making real the capacity to make the thing that defines humanity—the human genome," he pointed out. Endy and his colleague Laurie Zoloth suggested in an open letter to the conference organizers that synthesis of less controversial and more immediately useful genomes should be pursued instead.[3] The head of the original Human Genome Project, Francis Collins, also added a warning that this sort of genome synthesis project would "immediately raise numerous ethical and philosophical red flags."[4] Synthesizing a human genome appeared to many people to be an irresponsible use of science.

Although it may be the case that genomic self-synthesis by our species takes things to a new level, the practice of using technology to improve our native capacities is certainly not new. Humans have employed devices to enhance their biological and physiological abilities for millennia. If we can employ a technology or piece of design to ease a particular challenge, why not take the opportunity to improve the quality of our lives? From the first wooden legs affixed in Persia 2,500 years ago to the brain implants for stimulating specific neurons that are being developed today, we have become accustomed to a progressively more sophisticated blending of technology with the body.

For the more philosophically minded, connecting our physical selves more and more closely to assistive devices creates a new type of being. By now, the idea that the human body can be a site for the blending of the natural with the artificial is no longer unusual. When our bodies fuse with technology, we become a kind of hybrid composed of both biological and artifactual parts.

The idea of human and machine integration has created a new field known as *cyborg studies*. In many situations, it is accepted that we can live a higher quality of life if we are assisted by machine parts. Those parts can be heart pacemakers, computerized robotic limbs, or neuronal implants. But a cyborg does not have to be particularly complicated. The blend of biology and artifact can be something as simple as using a pair of reading glasses or a walking frame. The more seamless the interface between the human and the machine, the better it generally is for the human user. As time goes on and the complexity of the assistive device increases, clear lines between humans and their machine supplements are getting more and more fuzzy. Many within cyborg studies think this is a good thing. One of today's neuronal stimulation experts, Charles Lieber, has pointed out that one of the explicit goals of his work is "to blur the distinction between electronics as we know it and the computer inside our heads."[5]

The goal of synthesizing the human genome from scratch, however, is an entirely different sort of project. It goes far beyond simply creating a cyborg. Human genome synthesis does not seek to create a blend of the human and the artifactual. It seeks to remake the human from the inside out. Bill Clinton and Tony Blair pointed out at the conclusion of the first Human Genome Project in 2000 that our genome represents something special. It is understood by many to be our essence. As Endy and Zoloth claimed in their objection to the secret Harvard Medical School gathering, human genome synthesis could be used to "completely redefine the core of what now joins all of humanity together as a species."[6] This level of genetic remaking means that, even if we were genetically constructed with all the genes that belong in the human genome map, our species would have become something inherently different. We would be a product no longer of evolution but of technology. We would be "self-replicating" in a much more significant sense than anything that had occurred before. For the first time, we would possess genomes built by scientists. We would reproduce ourselves technologically rather than biologically. This would not just be in vitro fertilization. It would be in vitro creation.

The wisdom of making such a move is, to say the least, questionable. When Craig Venter succeeded in synthesizing a minimal bacterial genome in 2016, he had to admit that, even though he and his coworkers built the thing themselves in the laboratory, fully one-third of the genes the team had stitched together possessed an unknown function. Although this third was evidently essential for the bacterium to live, their builders confessed they had no idea what those genes did. The normally self-assured genomicist conceded that the process taught him that "we need to be a lot more humble about basic knowledge in biology."[7] A synthesized human genome six thousand times as long as the bacterium Venter's team constructed would contain a lot of DNA whose function would be a mystery. Humans would be remaking themselves from the gene up without being entirely sure what they were building. *Homo faber* would be taking an almighty big shot in the dark with its own genetic identity.

• • •

Hand-built genomes are not the only versions of human self-synthesis that loom over the Plastocene's horizon. If one candidate for the essence of every human being is DNA, another equally plausible candidate is the mind. The mind is perhaps an even more mysterious and elusive aspect of identity than DNA. The role the mind plays in creating our sense of self is incalculable. Attempts to synthesize the human mind lead us far into terra incognita. Although working with genomes ensures that any modifications made to our core at least keeps us located in the biological realm, working with consciousness provides no such guarantee.

A decade ago, an expert in nanotechnology and artificial intelligence named Ray Kurzweil wrote a book titled *The Singularity Is Near*. In this 650-page opus, Kurzweil explores a vision of the future that he predicts will be unleashed by the rapidly escalating powers of computers. The subtitle of the book reveals what the highly acclaimed futurist believes to be an inevitable consequence of this growing power—the ultimate transcending of human biology.

Kurzweil has established for himself an impressive track record of technological developments. In the 1970s, he was at the forefront of developing optical scanning devices to convert printed text into digital information. Soon afterwards, he built the first voice synthesizer for converting text into audible speech. Through a partnership with Stevie Wonder, he also invented the first keyboard synthesizer, earning in the process the National Medal of Technology from President Bill Clinton for inventions that change the life and culture of America.

In *The Singularity Is Near*, Kurzweil anticipates a future in which artificially intelligent machines gain a runaway intellect that exceeds anything the human brain can counter. He characterizes this time as the

"Singularity," drawing the name from a term used in physics to describe the point in a black hole where all known laws of physics cease to operate. Beyond the Singularity, all predictive bets are off. The Singularity represents, in Kurzweil's words, an event horizon "that is hard to see beyond." Such a runaway intelligence would be completely unfathomable to us and would usher in an entirely different world.

An early step on the way toward the Singularity would be the creation of a machine possessing computing power equal to that of the human brain. Kurzweil predicts that this will happen around 2020. Because the human brain is the most powerful machine in the biological world, crossing this threshold would have immense significance for evolution. Kurzweil describes it as "comparable in importance to the development of biology itself."[8] Whatever information-processing ability biology has managed to achieve over the three and a half billion years of evolution, technological society would now have surpassed it.[9]

Beyond that, Kurzweil sees a continued melding of biological brain power with nonbiological computational power. By 2029, Kurzweil predicts, all the functionality of the human brain, including its emotional dimensions, will be capable of being accurately modeled. If you know how to model them, it will not be long before those functions can be reproduced on machinery located outside of the brain. By 2045, Kurzweil predicts, the union of human intelligence with this still-expanding computing power will have exceeded the reach of our collective minds by a factor of a billion. This would mark the arrival of the Singularity.

Past the Singularity, the human future is unknowable. Kurzweil believes a "human-machine civilization" would be upon us that surpass our ability to comprehend it. One of the options available is the uploading of all the mental attributes taking place in a given brain to a separate "computational substrate." This means that our *minds* will be able to leave our *brains*. The inherent limitations of biology for thinking will no longer apply. Any limits that existed will be defined entirely by technology rather than by biochemistry. Humanity will become a completely different animal. In fact, *animal*, at that point, might no longer be the appropriate term.

With minds exportable from brains, we would move, according to documentarian James Barrett, "beyond the human era." When we become a "postbiological" species, the idea of a cyborg—an entity in which both the human and the artificial each play indispensable roles—becomes obsolete. Beyond the Singularity, the biological human would become increasingly dispensable. In the face of this shift, what it actually means to be human would be uncertain. Recognizing how disorienting this whole vision is, Kurzweil tried to calm some nerves by insisting that "future machines will be human, even if they are not biological." But it is unclear at this point what, if anything, would be left of our humanity in any recognizable sense. Technology would not only have allowed us to transcend the material limits of the physical world around us through technologies of deep manipulation like nanotechnology and synthetic biology. Technology would have enabled us to transcend our embodied selves.

If Kurzweil is right, the natural progression of genetics, nanotechnology, and computer science will move us inexorably toward a place where not only the world gets remade. We ourselves also would be remade. His predictions make it clear that there is a slippery slope between engaging such powers to transform the world and engaging them to transform ourselves.

Bill McKibben's desperate plea to draw a line in the sand with genetic technologies and to declare "Enough!" was motivated by the idea that we must do what is necessary to stay human. For some, there will no doubt be a strong desire to cross McKibben's threshold into a "posthuman"—or "transhuman"—world. Kurzweil himself has no qualms about making this move. Others side with McKibben and remain utterly repulsed by the whole idea of it.

Kurzweil's idea of the Singularity was intended to suggest a future that is so different from the present that it is literally unimaginable from where we currently stand. By Kurzweil's own definition, it is impossible to know what this future after the arrival of the Singularity might hold. Although we certainly cannot know everything that we might gain from the blending of human intelligence with computing power, it is possible to get a limited sense of what we will be giving up.

Four centuries ago, French philosopher René Descartes made the quaintly unscientific proposal that humans are a made up of a combination of two essential parts. He called these two parts mental substance and physical substance. "Cartesian dualism,"[10] as this two-substance view became known, was one of those ideas that became so embedded in common understanding that few people today suspect that it is an idea that owes its existence to certain powerful articulations of it at distinct points in intellectual history.

Descartes' proposition stuck because it seems to match so well what it actually feels like to be human. Our minds appear to us as some sort of immaterial essence that exists within our physical and biological bodies. In addition to these appearances, Descartes was not unaware of the compatibility of his view with his Christian faith. It is a belief in this separation of mind and body allows Christians, as well as a number of other religious faiths, to make sense of an afterlife.

This view is, however, starkly at odds with everything Darwin later bequeathed to us about evolutionary theory. The two-substance view articulated by Descartes crashes awkwardly against Darwin's suggestion that humans are entirely the products of a long and natural evolutionary process, descended like the rest of the biological world from common ancestors. Without a clear philosophical separation of mind and body, most atheists and agnostics suspect that as the body fades away at death, so does the mind fade with it.

The technologies that Kurzweil suggests are on the horizon could give new breath to Descartes' popular intuition. If we can survive as minds outside our bodies, the idea of an essential Darwinian union between the mental and the physical, between our conscious selves and our biological selves, will become unnecessary. In Kurzweil's world, it will not require a religious commitment to believe that a mind can transcend the death of the body. But if you are looking to keep the religious commitment, transhumanism may provide a twenty-first-century update to an age-old article of faith.[11] The downloading of consciousness into a computational substrate means that minds and bodies will have the potential to become dissociated. This might happen even before death, in fact whenever the holder of a mind chooses to dissociate them by downloading their consciousness. The long-held Cartesian intuition about humanity's true essence being disembodied might take on a new and contemporary significance.

• • •

Among the more reflective of the researchers working in molecular manufacturing, synthetic biology, and artificial intelligence, there is a growing sense that something fundamentally different is now at stake. The technologies fast approaching mark a different sort of change to the world. We are no longer altering surfaces to make life more congenial. We are changing deeply embedded elements of ourselves and our surroundings.

One of the scientists in the Sculpting Evolution lab at MIT has become convinced that these high stakes demand that the ways of conducting research into powerful emerging technologies must change. Synthetic biologist Kevin Esvelt believes that the range of techniques now at our fingertips across different research domains are so utterly transformative that we need a completely new process for engaging them. This new way would continually present the public with a clear account of what is at stake and a meaningful opportunity to say no. Science would be much more self-consciously directed toward the public interest. It would not be driven by hovering commercial interests and it would reject any attempt at secrecy enforced by patents or alliances with big companies and their market intentions.

For Esvelt, such views apply not least to the research he engages in himself. When discussing the potential release of a gene drive that could spread a terminal trait through a wild population of disease-carrying mice, Esvelt insists that "The only way to conduct an experiment that could wipe an entire species from the Earth is with complete transparency." The type of science that he advocates is completely open to the public ("All of it").[12] At every point in the project, the public should have the opportunity to say no. This call for a more open way of doing science has become Esvelt's personal mission, one that he shares with anyone who will listen. It is a mission that often puts him at odds with others working in similar fields.

Keekok Lee called them "deep technologies." Diane Ackerman called them "inventions that reinvent us." Whichever terms or phrases are chosen, as we flirt more and more with the different elements of the Synthetic Age, a drastically different future awaits. It is a time likely to be filled with major shifts in the reality we experience. Due to the magnitude of these shifts, it seems critical to look hard at these transformations and decide collectively if, and in what ways, we wish to embrace them. The changes that await are far too profound to be left entirely in the hands of technological visionaries and the cluster of economic interests that are constantly stalking them. Unless we make a conscious decision to go down certain paths and to avoid others, the Synthetic Age ahead will host not just a radical reengineering of the world that surrounds us. It also will host a dramatic reengineering of ourselves.

For some, this is a future to get excited about. Nothing, after all, ever stays the same. But these changes also could be disorienting enough to leave us isolated and utterly confused while drifting across an unfamiliar and unknowable new reality. It is a path that, at the very least, we should not let ourselves be pulled down unwittingly.

The work being pursued by people like Kurzweil and Esvelt reveals that technologies at the dawn of the Plastocene are powerful enough that they demand from us unprecedented ethical scrutiny. If there was ever a time when it was important to think hard about nature and its relationship to technology, that time is now. If there was ever a time to reflect deeply on who gets to make the decisions about implementing these changes, that time is upon us. The plea for a more democratic approach to how we choose the future we will inhabit may end up being one of the primary demands of the Synthetic Age.

11 The Transitional Moment

The recent explosion of interest in the idea of a changing epoch can be pinned fairly precisely to the publication of a single essay. Paul Crutzen already had the Nobel Prize to his name when in 2000 he and a marine ecologist named Eugene Stoermer first made their claim about collective human impacts causing an exit from the Holocene. Reflecting on the degree to which humans had changed the earth and its biological as well as geophysical systems, Crutzen and Stoermer concluded that "it seems to us more than appropriate to emphasize the central role of mankind in geology and ecology by proposing to use the term 'Anthropocene' for the current geological epoch."[1]

The brief paper in which they published these ideas, written for an obscure academic newsletter, marked the beginning of a radical shift in our species' self-image. The planet was not as vast as we originally thought. It could be entirely transformed by our actions.

Part of the reason Crutzen and Stoermer's essay left such a mark was that, by the turn of the millennium, the public had become increasingly attuned to the phenomenon of human-caused planetary change. Several decades of continuous environmental messaging had driven home the idea that our species was making an almighty mess of its home. Climate change had become a growing global concern. The suggestion that the earth was experiencing a sixth mass extinction was well-established in the guilty corners of the public mind. The ability to find haunting images on the Internet of extinct species such as the Javan tiger and the passenger pigeon had made the finality of environmental destruction palpable to most people. Threats to surviving species such as rhinos and polar bears had created a whole generation of school children for whom slogans about saving the whales and protecting the rainforest had literally been served with their lunch.

Crutzen and Stoermer's opinion piece was an attempt to capture a sense of just how large the human footprint on the earth and its systems had become. They drew attention to a familiar laundry list of impacts on the biosphere. This litany included the scale of diversion of fresh water for human uses, the exponential increase in the human population, the amount of atmospheric nitrogen being fixed for agriculture, the extent of the destruction of coastal mangrove forests, the explosion in the number of farm animals roaming earth's pastureland, and the quantities of carbon and sulfur dioxides being belched into the atmosphere from the burning of fossil fuels. They also pointed to the sheer extent of the planet's surface—well over 50 percent—now converted primarily to the satisfaction of human needs.

What these two senior statesmen thought particularly significant was how these human impacts were occurring on scales that dwarfed comparable natural processes. In the natural world, for example, nitrogen is continuously drawn out of the air by leguminous plants such as peas and beans with the help of trillions of bacteria. But the nitrogen captured industrially through the Haber-Bosch process—in excess of a hundred million tons per year—moves more of this element out of the atmosphere and into the ground on an annual basis than all of the natural bacterial processes combined.

In a similar vein, the movement of soils and rocks through the mechanical forces present in agriculture, industry, and urbanization now exceeded the movement of soils and rocks through erosion. The total mass of water stored behind the dams that replumb the world's rivers and streams literally changed the planet's spin. Species were going extinct due to human activities at a thousand times the background rate suggested by the fossil record. And in one of the most noted calling-cards of the planetary transition, humans had put more carbon into the atmosphere than natural processes had done for at least 800,000 and perhaps as many as three million years. The way nature had always operated was now looking increasingly quaint and inconsequential when compared to the feats of planetary engineering achieved by the earth's rambunctious hominids.

To press their case in the language that mattered for epoch naming, Crutzen and Stoermer imagined a future geologist looking back and investigating this period of Earth's history. Evidence located in sediments and rocks tells stratigraphers a story about the major planetary shifts occurring at any one time. Rapid fluctuations in temperature or in the concentration of atmospheric gases, explosions in certain types of plants or pollens, changes in the ocean biota, and even tell-tale deposits left by asteroid strikes can all be identified and dated by digging carefully through the layers that are found beneath our feet.

Crutzen and Stoermer imagined a future stratigrapher sifting through the sediments deposited during the current age and concluded that the markers left by human works would be the most defining features they would find. Sediments would tell of planetary-scale rearrangements of earth and water. Fossil records would reveal startling rates of species extinction. Examination of rocks would reveal a range of entirely new "technofossils" made up of plastics and other man-made substances. Drilled ice cores would reveal the rapid increase in carbon dioxide in the ambient air. The empirical signals would all confirm that this slice of history represented a planet shaped by people.

Crutzen and Stoermer were not saying anything completely new when they made the case for epochal change. There had been several previous attempts to articulate something like the same idea. A nineteenth-century Italian priest with an interest in fossils named Antonio Stoppani—later a professor of geology at the University in Milan—used the phrase "the anthropozoic era" to capture the extent of the human-induced changes he observed around him. Stoppani painted a poetic picture of humanity as "a new telluric force that for its strength and universality does not pale in the face of the greatest forces of the globe."[2]

An American of roughly the same time period, Thomas Chrowder Chamberlin, used the term "the psychozoic era" to capture a similar idea. "Man is the most important organic agency yet introduced into geological history," waxed Chamberlin, showing just a touch less poetry than his Italian counterpart: "Both the organic and inorganic agencies of geological progress are powerfully influenced by him."[3] The Russian geologist Alexei Pavlov—not to be confused with the famous Pavlov of drooling dogs fame—may have been the first to coin the term *Anthropocene* in 1922. But in the absence of an ecologically aware audience and with the planet still appearing to be unfathomably large, none of these early figures was able to make his case in a way that the idea of a human-directed epoch could really establish itself in the public consciousness.

A few generations later, a contemporary writer started promoting a modern version of the same idea. The environmental movement and the United Nations' Kyoto Protocol on climate change were on everyone's mind when *New York Times* columnist Andy Revkin[4] started using the term *anthrocene* in the mid-1990s to capture the new zeitgeist. Like McKibben and his idea of "the end of nature," Revkin knew something big was going on. But the less melodic sound of Revkin's chosen term—and also perhaps the lack of a Nobel Prize to his name—meant that he too failed to popularize the radical geological idea.

Aided perhaps by a turn-of-the-millennium reflectiveness about history, it was Crutzen and Stoermer's essay that finally succeeded in launching the idea of a human-dominated epoch. Their idea of the Anthropocene took off not because people cared particularly about geology. It took off because people cared about what a "human age" signified. It made a massive statement about human power that, love it or hate it, many people found hard to resist. The idea that our species could be geologically—or even astrally—significant tapped into some deep psychological well. The term quickly escaped the bounds of academic conferences and journals with names like *Nature Geoscience* and the *Journal of Geophysical Research* and took on a vigorous popular life of its own. Articles in *Time*, *National Geographic*, and *The Economist* brought the term out of the university setting and into the wider culture. People still may not have cared much about rocks. But they cared about the promise of being able to shape a whole planet.

As a result of the term's popular rise, the world's most eminent geologists have been considering whether to make Crutzen and Stoermer's neologism official. The people who would formalize the renaming are the men and women of the International Union of Geological Sciences, people whom environmental writer Robert MacFarlane has called "the monks and philosophers of the earth sciences" for the gravity of the work they do. The Union will base its decision largely on the advice of a subgroup named the International Commission on Stratigraphy.

For more than two years, the International Commission tasked a few dozen suitably qualified researchers to survey a mountain of evidence from climate science, biology, hydrology, the geosciences, paleontology, and other disciplines in order to assess whether the naming of a new epoch was warranted. In an article published in *Science Magazine* in January 2016, these researchers drew the preliminary conclusion that the earth had indeed both "functionally" and "statigraphically" left the Holocene and entered the Anthropocene. Over the next few years, the commission will decide whether to accept the working group's recommendation and pass it up to the next level, where the International Union of Geological Sciences would decide whether to certify the designation.

It may take a while. When things are measured on geological time scales, not much official business can be considered terribly pressing. Nevertheless, the wheels that would formalize the renaming are in motion. Given how rarely it occurs, a transition to a new planetary epoch would be monumental. It would dwarf the significance of the recent new millennium. Those historical waypoints occur predictably every thousand years. New epochs come around at highly irregular intervals every few million.

Such a designation would also be a little odd. No other epoch has been named at the very moment it was beginning. In fact, of all the previous epochs, only the Holocene was named when it was still going on. At the time it was named, the Holocene epoch was already more than 11,500 years old. Although the working group from the International Commission on Stratigraphy have made a compelling case that a threshold has been crossed and that the Holocene is now behind us, there is currently a great deal of confusion about the best way to proceed. To some people, branding the emerging epoch with our own name when it has only just started seems like an act of gross geological conceit. Others who are more accustomed to the practice of epoch naming ask why there should be such a big rush to decide any of this.

The whole discussion about the shifting epoch seems to have got off on the wrong foot. The fact that our species has accidentally left its signature in every remote bay, on every mountain top, and across every continent is certainly a major cause for introspection. But it does not seem like the right opportunity to celebrate our untidiness by naming the next epoch in our honor. A lot remains to be determined about the contours of millennia that lie ahead. At the very beginning of this new epoch, we know little about the shape it will or should take.

What we can know is that from this point onward, a certain subset of the human population will have at their fingertips some extraordinary powers for remaking the natural world. For the first time, people will be able to take what nature has been doing by itself for billions of years and start doing it themselves. Climate, ecology, and molecular biology may increasingly be replaced with synthetic versions of themselves. Earth's most formative processes may become more and more human-directed.

Climate engineer David Keith, perhaps mulling over some of the Heideggerian philosophy he learned from Albert Borgmann in Montana, notes the moment of history we occupy dispassionately: "About a million years after inventing stone-cutting tools, ten thousand years after agriculture, a century after the Wright Brothers' flight, humanity's instinct for collaborative tool building has brought us the ability to manipulate our own genome and the planet's climate."[5]

Keith feels keenly the weight of the powers unleashed by synthetic biology, climate engineering, and similar technological advances. The question he leaves unanswered is whether we should aggressively take up these new powers and become more and more deeply involved in remaking ourselves and the world. If Keith appears to endorse a brazen form of planetary management due to his advocacy of climate engineering research, he does so with great reluctance. His ski trips into Arctic wildernesses still mean too much to him. The hesitant climate engineer admits that he still longs for the idea of a natural world that lies beyond human reach.

It is easy to see why one might be tempted to side with the hands-on approach advocated by Crutzen and his cohort. As a species, *Homo sapiens* are by their nature doers and fixers. We are on the hook for causing a great deal of disruption to the planetary system. Through a suite of new technologies, we might now have the potential to repair some of the damage, even if this means recalibrating several of the earth's most essential metabolic functions. If done smartly, engineers and ecosystem managers might subtly rewire the planet to make it more resilient to our excesses. In the process, we might be able to engineer our way around previously firm ecological limits and reverse harms that we thought were permanent. With planetary systems rejuvenated and made more resilient, a more optimistic future might await. The environment might be less vulnerable. Economic growth might be less constrained. A few decades into this new reality, says ecomodernist and supporter of climate engineering research Jane Long, we will have all learned to find beauty in the managed and cultivated world we have created. It is often the case that we come to love the objects for which we truly practice our care.

Long, Crutzen, and other ecomoderns are convinced that these steps are not only appropriate but also inevitable. We now live on a different planet with different rules of engagement. It is imperative that we wake up to this. Crutzen thinks our sleepy acceptance of old ways of thinking is regrettable. "It's a pity we're still living officially in an age called the Holocene," he has written.[6] It would be better for humanity to acknowledge the changing epoch and to start playing a different game.

There is no doubt some truth to the claim that we need a different and more self-aware game. Things *are* different now. But numerous voices allied with McKibben's do not share Crutzen's particular vision of what this more self-aware game entails. Many think that, at the very time we are acknowledging the extent of our impact, ramping up our interventions into the natural order is a big mistake.

"The disintegration of what is natural into what is artificial, and the consequences of this erosion are beyond sobering," says Montana nature writer Rick Bass.[7] When taking an aggressively interventionist approach, there is increasingly little around us that we must accept for what it is. The physical and biological world becomes more and more contingent, ready to be remade at our whim. More than ever before, it becomes *our world,* and we assume total responsibility for shaping its future and our future in it.

No doubt this is part of what Bass finds sobering: there is nowhere else to look, nobody else to blame, just our own imperfect decisions about what is best. Jason Mark worries that a world increasingly transformed through technology becomes more and more like a hall of mirrors, where we see reflections of ourselves everywhere we look. He calls this urge to transform everything "species narcissism on a planetary scale." Without the counterweight of an independent nature that can offer resistance to our desires, we risk driving ourselves insane.

Bass, Mark, and others who are rejecting the urge to ramp up our interventions are also concerned that we might misjudge how much certainty and control we can have over the world we will make. Despite assuming godlike powers, we should recognize that omnipotence and omniscience have never been our strong suit. Capricious forces still reside deeply within biology, geology, and the slow unfolding of planetary history. We might be lulled into forgetting the inherent wildness that still resides on a dynamic, living planet. Our response to these forces have long been twofold. They are not just forces that we should treat with caution. They are forces we should deeply admire.

In the early 1990s, a private foundation embarked on an audacious experiment to simulate a fully functioning ecological system by constructing a facility it called "Biosphere 2" in the Arizona desert. The intention was to create an entirely self-contained, biological life-support system that could maintain a small number of human voyagers for a period of two years. Its name was a reflection of the attempt to recreate an ecology that closely mimicked the earth's own. The facility was built using the best technology and research science available. On a very small scale, the experiment was perhaps our species' closest previous attempt at synthesizing this planet.

Although some interesting lessons were learned, Biosphere 2 is widely thought to have been an embarrassing failure. The inability of the prospective "bionauts" to create an inhabitable world was as much a consequence of human dysfunction as it was a consequence of the considerable failures in structural and ecological design. There was too much that the designers of Biosphere 2 did not know about the ecology they had built. There was too much that they did not anticipate about the social dynamics of a human crew surrounded by an entirely constructed environment.

Biosphere 2 might be considered a cautionary parable for the Synthetic Age. Although significant human influence on the earth's future is now inevitable, nothing will guarantee that even the most conscious and well thought-through attempts at earth synthesis will work out as imagined. The unpredictability that remains present in both natural and cultural systems ensures that no such guarantee can be offered. Neither biology nor society is likely to display obedience to our designs for very long.

Some of the red flags raised by the projects under consideration are obvious. Setting self-nourishing and self-replicating machines or organisms free in the environment to perform work for us seems inadvisable. Designing genomes that have the chance to mutate on us is a mighty gamble, especially when conceding the ignorance we still possess about the relationship between the genome and the microbiome. Attempting to manage physical systems as large and as chaotic as the planet's climate not only is inherently hazardous but also smacks of overconfidence. Setting in motion a suite of hard-to-reverse biological and ecological processes beyond our watchful eyes creates the distinct possibility of a Synthetic Age that will turn viciously on us.

There are also important questions about the magnificence and wonder of the natural world that should give us pause. The complexity and beauty of the world admired by Muir, Leopold, and countless other environmental thinkers is the direct consequence of a long and unpredictable evolutionary odyssey. This odyssey was not engineered or designed. It simply happened, taking place through a remarkable concatenation of events steered mostly by luck and by chance. As part of this unfolding of history, cataclysmic events occurred. Many were very painful and wrought great havoc. Some of these events are still likely to occur, even when genomes, ecosystems, and climates are being synthesized by well-meaning technicians.

These biological and geological realities mean that we need to think hard about where to go from this transitional moment. Humanity's outsized impact means that our responsibility for the earth has grown. There is no doubt that we will be making decisions that will shape the earth and its ecology over the coming epoch. But big choices of direction remain. One possible future is a Synthetic Age that takes disrupted planetary processes by the scruff of the neck and entirely remakes them along the lines our engineers think will work better. Another is a humbler epoch that mixes careful innovation in some areas with repair of Holocene baselines in others. Both have their merits, but we should be wary of the lures that will pull us too quickly in one direction or another. We should be conscious of who is making these decisions for us and where their interests lie.

Political history reveals just how much people can be angered by the idea that their future is being decided for them by far-off elites. Whatever callous manipulations of fact and woeful distortions of truth characterized both the Brexit referendum and the Donald Trump presidential campaign in 2016, both of these political movements succeeded by suggesting that people out there, in Brussels or in big New York City banks, were deciding our future in ways that served their own interests and not ours. A majority of voters decided that this was a fundamental injustice in need of correction.

With so much at stake in a Synthetic Age, a parallel concern could be in play. The approach to science recommended by Esvelt concedes that, when the stakes are high, people should not have their futures—and the future of the environments that surround them—decided on their behalf. Stakeholders should be given the opportunity to know what is coming their way and be offered meaningful input on whether they actually want that type of a future. If the extent of this input is simply an after-the-fact choice about whether to purchase a particular end-product, then too much has already been decided. Too little has been shared about what is happening to their world.

At the dawn of a Synthetic Age, the future of nature should not be determined simply by what is possible. *Can* has never automatically entailed *should*. The shape of the future must involve deliberation and discussion by as many people as possible. Some of these people will be highly qualified experts with relevant technical knowledge. Many more will be teachers and parents, workers and retirees, young people and those representing the interests of generations who will be born in the future that unfolds. As Jedediah Purdy warned, it would be better not to fall into the future through drift and inadvertence. We must learn what we can about the technologies headed our way and participate vigorously in the debates over what shapes they will take. The future must be—as much as it can—a matter of deliberate and considered choice.

Making big choices is always hard. Making irrevocable choices for the whole planet is unprecedented. But at this point, we have changed too much to stand back and do nothing. We need to look at as many of the various options as we can, talk about them, argue about them, investigate and research them as thoroughly as possible. Conducting this discussion thoughtfully, fairly, and inclusively is perhaps the worthiest, and certainly the most important, political task of our time. It is also one that we can no longer shirk.

Although the prospect of such a complex discussion about the shape of the Synthetic Age is no doubt intimidating, none of it should be a cause for despair. After all, the ability to think through and contest with fellow members of our clan the options that are in front of us is our singular gift. It is part of both the burden and the joy of being *Homo sapiens*, the wise species.

Afterword: A Postscript on the Wild

On August 7, 2015, the body of a hiker was found about half a mile from the Elephant Back Loop Trail in Yellowstone National Park. The Park Service announced that the hiker had been mauled and partially eaten by a grizzly bear. A hunt for the responsible bear quickly yielded a mother and two cubs loitering in the area. The mother was trapped, and after DNA evidence revealed that she was responsible for the hiker's death, she was euthanized. The two cubs were moved away from Yellowstone to live the rest of their lives in an Ohio zoo.

The hiker attacked by the bear, Lance Crosby, was an employee of one of the medical clinics located in the park. He had worked in the park for five summers and was familiar with the country and the risks it contained. Crosby was well liked by his coworkers. At the time of the attack, it appears he might have been going for a quick hike to test out the strength of an ankle he had injured the previous week. Friends told Park Service authorities that Crosby often hiked alone and never carried bear spray. Even though he would have known that this was not recommended in the park, Crosby had plenty of experience in Yellowstone and felt like he knew what to look out for.

Crosby's wife reported that her husband loved the Yellowstone landscape and had always fostered a deep interest in bears. Because of his interest in natural history, Crosby no doubt understood some of the evidence that Yellowstone was in the process of becoming a different landscape. He knew that climate change had altered the seasonal rhythms of the park and was beginning to create shifts in some of the park's vegetation. He had experienced the unusually early beginnings of the summer tourist season and had worried about the increasing risks of wildfire during late summer and fall.

Crosby also appreciated that the park was in many respects a carefully constructed landscape, with Bannock and Shoshone Indians being forced to make way at the Park's creation in 1872. He knew that park biologists were busy removing nonnative trout from Yellowstone Lake. He was aware that the bison were being intensively managed during the winter months through hunting and culling to diminish the risk of transferring brucellosis to Montana's cattle herds. He had seen wolves in the park wearing bulky radio collars so they could be studied by an endless parade of ecologists and wildlife biologists. He had also no doubt watched Park Service employees towing around the giant culvert traps used to capture and relocate problem bears from various high traffic areas.

A great deal of hands-on management—"gardening," as Marris calls it—goes into keeping up the appearance of Yellowstone Park in the manner that its visitors have come to expect. If Crosby had read any writings by Emma Marris or Gaia Vince, he might have been tempted to think of the beautiful landscape in which he spent his last five summers as postnatural or postwild. Certainly, he would have known that today's park in its heavily manipulated form lacked the naturalness it possessed ten thousand or even a hundred fifty years ago.

Yet when that sow bear came to within a few feet of him, Crosby probably understood for a terrifying few seconds that Yellowstone was far from postwild—not now and not ever. Many of the processes that gave the ancient caldera and its ecologies their shape remain present and operational. Blizzards still rake the landscape in winter. Fires still burn ferociously in summer. Evolutionary pressures still operate on the biota. Photosynthesis and respiration continue without pause. Predation is still present, and defensive behaviors are still passed on between generations of the park's fauna. The bear that attacked Crosby was still driven by powerful urges that had been fine-tuned by its species' fifty thousand years of inhabitation of the North American continent. These are urges that no practices or interventions by wildlife biologists or park managers have any chance of quelling. Wildness, in other words, retains its place in Yellowstone, still lurking in the cracks of an increasingly managed system.

Wildness is, in fact, the riddle that will inhabit every element of a synthetic future. It will continue to reside not only in ecological landscapes and in the predators they contain but also in every practice and technology that we will try to develop. It will be found in the nanobots that Drexler was worried might run out of control and convert the earth into a grey goo. It will exist in the synthetic organisms that Venter recognizes must be prevented from escaping from the lab or from turning pathogenic. It will continue to course through the veins of the species optimistically relocated whose number might come up in the game of ecological roulette being played by ecosystem managers. It will burst forth from more intense monsoons that will unexpectedly shift five hundred miles east and come a month later as the unanticipated result of a well-intentioned, but misguided, attempt at solar radiation management. It will prowl within any human genome that is synthesized in the lab. Every technology and practice will contain important traces of wildness that will remain callously indifferent to our plans and our desires.

Wildness will continue not only as a property of the technologies we build; it will persist as a property of the builders themselves. As spontaneous social and biological beings constantly evolving new patterns of behavior in response to changing circumstances, both individuals and societies will be eternally in wildness's thrall. Swirling and unpredictable throngs will rapidly coalesce around charismatic individuals. Extensive cultural behaviors will take unexpected turns, whether in the form of radical political movements, the rapid adoption of a new technology, or the scourge of fundamentalism. An elderly woman, years into a routine of walking to a local store, will abruptly turn left instead of right. The spontaneity within us all will continue to produce both spectacular human successes and terrifying political and economic failures in ways that cannot be anticipated.

Wildness, then, is a perpetually mixed blessing. On the one hand, it ensures that the beauty, the spontaneity, and the enchanting unpredictability of the world outside of our grasp will always exist alongside our inventions. In relentlessly evolving species and ecologies, in the lotteries constantly won and lost between predator and prey, in unexpected downpours and luminescent rainbows, and in the unceasing physical and thermodynamic forces that have always shaped the home planet, wildness will ensure that there will always be mystery and wonder to behold in whatever sort of Plastocene we choose to create. The autonomy and indifference to our goals that wild animals and wild landscapes display will remain vital for keeping our projects and our dreams in perspective.

However, there is another side to this wildness that it would be foolish to forget. In its fickleness, its unpredictability, and its capacity continually to exceed our expectations, wildness will ensure that remaking the earth will always remain a game of high chance. When we insert ourselves so deeply into the workings of a planet, we are unlikely to be able to predict all of the consequences of our actions. There are serious risks to letting ourselves be seduced by the sublime beauties of technology.

The gears of geological epoch naming are already turning, and before long, stratigraphers may decide to rename our time "the human age." If that happens, we might take that opportunity to inhale deeply, survey what lies around us, and reflect. The renaming will say something important to us about who we are and what we might become. But in that moment of reflection, our species would do well to hesitate for as long as possible before moving ahead. The pause will offer a chance to take on board the fact that, despite our best intentions, nature and the billions of fast-changing lives it contains are not likely to lay down and entirely do our bidding. Not even after the monks and philosophers of the earth sciences have named the next epoch our own.

Notes

Introduction

1. A gaff is a wooden or metal club with a steel hook embedded in its end that is designed to help fishermen haul big fish over the side of a fishing boat.

2. The prefix *anthropo-* is derived from a Greek word for "human."

3. Paul Crutzen with Christian Schwägerl, "Living in the Anthropocene: Toward a New Global Ethos," *YaleEnvironment360*, January 14, 2011, http://e360.yale.edu/features/living_in_the_anthropocene_toward_a_new_global_ethos.

4. Throughout this book, I use the terms *Synthetic Age* and *Plastocene* interchangeably. Both terms suggest that a world that was once the product of natural processes increasingly is becoming something we deliberately construct.

5. A further reading section at the end of the book points toward some of the sources of the ideas described. Endnotes and citations are kept to a minimum.

Chapter 1

1. Feynman, "There's Plenty of Room at the Bottom."

2. It is still not possible to "see" anything at the atomic scale because the wavelength of the light we use to see is much greater than the diameter of an atom. A scanning tunnel microscope can, however, create a visual representation of what is going on down there by using a current to provide an electrical representation of what an atomic surface or arrangement "looks" like.

3. Quoted in Joachim Schummer and Davis Baird, eds., *Nanotechnology Challenges: Implications for Philosophy, Ethics and Society* (Singapore: World Scientific, 2006), 421.

4. Banana Boat also reassured advocates who might take a stand against cruelty to innocent fruits that its products do not contain any bananas.

5. Mark R. Miller, Jennifer B. Raftis, Jeremy P. Langrish, Steven G. McLean, Pawitrabhorn Samutrtai, et al., "Inhaled Nanoparticles Accumulate at Sites of Vascular Disease," *ACS Nano* 11, no. 5 (2017): 4542–4552.

6. U.S. Environmental Protection Agency, "Chemical Substances When Manufactured or Processed as Nanoscale Materials: TSCA Reporting and Recordkeeping Requirements," 2017, https://www.regulations.gov/document?D=EPA-HQ-OPPT-2010-0572-0137.

Chapter 2

1. A striking picture of this first piece of molecular manufacturing is widely available online and is worth looking at.

2. Sündüs Erbaş-Çakmak, David A. Leigh, Charlie T. McTernan, and Alina L. Nussbaumer, "Artificial Molecular Machines," *Chemical Reviews* 115, no 18 (2015): 10157.

3. Drexler, *Engines of Creation*.

4. Open letters between Drexler and Smalley published in *Chemical and Engineering News* 81, no. 48: 37–42, http://pubs.acs.org/cen/coverstory/8148/8148counterpoint.html.

Chapter 3

1. A final and even more precise version of the human genome map was released in 2003, after which the Human Genome Project was formally declared finished.

Chapter 4

1. Quoted by Andrew Pollock in "His Corporate Strategy: The Scientific Method," *New York Times*, September 4, 2010, http://www.nytimes.com/2010/09/05/business/05venter.html.

2. Bill Joy, "Why the Future Does Not Need Us," *Wired* 8, no. 4 (April 2000), https://www.wired.com/2000/04/joy-2.

3. After faltering technological progress and a fall in oil prices in 2014, Exxon scaled back its investment in synthetic biofuel technology. A 2017 breakthrough has reignited the company's enthusiasm. See Imad Ajjawi, John Verruto, Moena Aqui, Leah B. Soriaga, et al., "Lipid Production in *Nannochloropsis gaditana* Is Doubled by Decreasing Expression of a Single Transcriptional Regulator," *Nature Biotechnology* 35, no. 7 (2017): 645–652.

4. J. Craig Venter Institute press release, "First Self-Replicating Synthetic Cell," May 20, 2010, http://www.jcvi.org/cms/press/press-releases/full-text/article/first-self-replicating-synthetic-bacterial-cell-constructed-by-j-craig-venter-institute-researcher.

5. McKibben, *The End of Nature*, 213–214.

6. Paul Crutzen, "The Geology of Mankind," *Nature* 415 (January 2002): 23.

Chapter 5

1. Leopold, "Marshland Elegy," in *A Sand County Almanac*.

2. Leopold, "The Outlook," in *A Sand County Almanac*.

3. An exchange at the Aspen Environmental Forum (2012) recounted at http://grist.org/living/save-the-median-strip-or-how-to-annoy-e-o-wilson.

4. Emma Marris, "Handle with Care," *Orion Magazine*, May/June 2015, https://orionmagazine.org/article/handle-with-care.

5. Marris, *Rambunctious Garden*.

6. Ellis, "The Planet of No Return"; and Erle C. Ellis, "Too Big for Nature," in *After Preservation: Saving American Nature in the Age of Humans*, ed. Ben Minteer and Stephen Pyne (Chicago: University of Chicago Press, 2015), 26.

7. Federal Ministry for the Environment, Nature Conservation, and Nuclear Safety, "National Strategy on Biological Diversity," 2007, http://www.bmub.bund.de/fileadmin/bmu-import/files/english/pdf/application/pdf/broschuere_biolog_vielfalt_strategie_en_bf.pdf.

8. Stephen Jay Gould, *Time's Arrow, Time's Cycle: Myth and Metaphor in the Discovery of Geological Time* (Cambridge, MA: Harvard University Press, 1987), 3.

Chapter 6

1. Due to the salvational nature of the technique, some have suggested calling it the biblical-sounding *directed diaspora*.

2. Stephen G. Willis, Jane K. Hill, Chris D. Thomas, David B. Roy, Richard Fox, David S. Blakeley, and Brian Huntley, "Assisted Colonization in a Changing Climate: A Test-Study Using two U.K. Butterflies," *Conservation Letters* 2, no. 1 (2009): 46–52. DOI: 10.1111/j.1755-263X.2008.00043.x.

3. Robert Elliott, *Faking Nature: The Ethics of Environmental Restoration* (New York: Routledge, 1997).

4. Fred Pearce argues that many of these attempts to correct "mistakes" are both a waste of money and ecologically unnecessary.

5. In 2017 the first human embryos were edited using the CRISPR technology. Hong Ma, Nuria Marti-Gutierrez, Sang-Wook Park, Jun Wu, et al., "Correction of a Pathogenic Gene Mutation in Human Embryos," *Nature* 548 (August 2, 2017), doi:10.1038/nature23305.

6. Emma Marris, "Humility in the Anthropocene," in *After Preservation: Saving American Nature in the Age of Humans*, ed. Ben A. Minteer and Stephen J. Pyne (Chicago: University of Chicago Press, 2015), 48.

7. Anticipating a future need for the DNA of today's vanishing species, several organizations are collecting the genomes of existing species in the hope that, at some point in the future, these samples might prove useful. Examples include the Frozen Ark project in Nottingham, UK, and the Frozen Zoo at the San Diego Zoo's Institute for Conservation Research.

8. If the idea of bringing an extinct species back from the dead sends chills up your spine, consider that similar technologies could be used to create cross-species clones of living but highly endangered species like the Javan rhinoceros. DNA from surviving Javan rhinos could be inserted into the eggs of close but less endangered relatives like Sumatran rhinos. People might think it better to have a cloned Javan rhino walking through the Indonesian jungle born of a Sumatran rhino mother than to have no Javan rhinos at all.

9. Stewart Brand, "The Dawn of De-extinction: Are You Ready?," TED Talk, https://www.ted.com/talks/stewart_brand_the_dawn_of_de_extinction_are_you_ready/transcript?language=en.

10. Quoted in "Mammoth Genome Sequence Completed," BBC News, April 25, 2015, http://www.bbc.com/news/science-environment-32432693.

11. Scott R. Sanders, "Kinship and Kindness," *Orion Magazine*, May/June 2016, 34.

12. Kevin Esvelt, George Church, and Jeantine Lunshof, "'Gene Drives' and CRISPR Could Revolutionize Ecosystem Management," *Scientific American*, July 17, 2014, https://blogs.scientificamerican.com/guest-blog/gene-drives-and-crispr-could-revolutionize-ecosystem-management.

13. Sanders, "Kinship and Kindness."

Chapter 7

1. International Agency for Research on Cancer, press release no. 180, December 5, 2007, https://www.iarc.fr/en/media-centre/pr/2007/pr180.html.

Chapter 8

1. Alvin Weinberg, *Reflections on Big Science* (Cambridge, MA: MIT Press, 1967).

2. One problem with technical fixes like antilock brakes is that they can give people a false sense of security. People usually react to that changed sense of security by driving faster, hence eradicating some of the benefits that should have been gained.

3. This field is also called *geoengineering* and sometimes *climate remediation*, although *climate engineering* has recently become the preferred term.

4. Paul J. Crutzen, "Albedo Enhancement by Stratospheric Aerosol Injection: A Contribution to Resolve a Policy Dilemma?," *Climatic Change* 77 (2006): 211–219. DOI: 10.1007/s10584-006-9101-y.

5. The oceans have already given us a temporary reprieve from considerable warming by naturally absorbing 30 to 40 percent of the carbon dioxide humans have emitted as well as up to 90 percent of the heat those remaining greenhouse gases are trapping.

6. At the end of the book *The End of Nature*, in an act of defiance, McKibben declares that he refuses to accept the "clanging finality" of the position he just spent a couple of hundred pages defending. He decides to throw himself into fighting climate change, "hoping against hope" that this irreversible loss can be prevented.

7. Paul J. Crutzen, "Geology of Mankind" *Nature* 415 (January 3, 2002): 23.

Chapter 9

1. Another reason that the prospect of spraying any kind of aerosol into the sky is sure to generate a hostile reception is the so-called chemtrail conspiracy theory. Chemtrail conspiracy theorists worry that some nefarious government power is already spraying chemicals out of commercial airplanes in order to exert control over the unsuspecting public below. Proposing to do something similar as a response to climate change stirs suspicion.

2. Royal Society, "Geoengineering the Climate: Science, Governance, and Uncertainty," 2009, https://royalsociety.org/~/media/Royal_Society_Content/policy/publications/2009/8693.pdf.

3. Another global climate impact connected to the rear ends of mammals is the suspected massive reduction of methane emitted into the atmosphere by the overhunting of herbivores in the late Pleistocene and Holocene. Although these extinctions and the attendant reductions in methane emissions could have caused some temporary global cooling, any trend in that direction has been reversed by the addition of billions of highly flatulent domesticated animals to satisfy humanity's hunger for meat.

4. Naomi Klein, "Geoengineering: Testing the Waters" *New York Times*, October 27, 2012, http://www.nytimes.com/2012/10/28/opinion/sunday/geoengineering-testing-the-waters.html.

5. Sabine Fuss, Josep G. Canadell, Glen P. Peters, Massimo Tavoni, et al., "Betting on Negative Emissions," *Nature Climate Change* 4, no. 10 (2014): 850–853.

6. Massimo Tavoni and Robert Socolow, "Modeling Meets Science and Technology: An Introduction to a Special Issue on Negative Emissions," *Climatic Change* 118 (2013): 13.

7. David Keith's Carbon Engineering made the shortlist of companies being considered for this prize.

8. Shortly after the end of the Vietnam War, an international treaty known as ENMOD (the Convention on the Prohibition of Military or Any Other Hostile Use of Environmental Modification Techniques) banned the use of weather modification as a military strategy. Some commentators have suggested that the sentiment behind ENMOD should also apply to proposed efforts at climate engineering.

9. Jason Mark, "Hacking the Sky," *Earth Island Journal* (Autumn 2013), http://www.earthisland.org/journal/index.php/eij/article/hacking_the_sky.

10. This phrase was first used by Dipesh Chakrabarty in "The Climate of History: Four Theses," *Critical Inquiry* 35, no. 2 (2009): 197–222.

Chapter 10

1. George Whitesides, "The Once and Future Nanomachine," *Scientific American*, September 16, 2001, 75.

2. They later renamed the project "Human Genome Project—Write" in order to contrast it with the project completed during the Clinton presidency, which they referred to as "Human Genome Project—Read."

3. Quoted in Joel Achenbach, "After Secret Harvard Meeting, Scientists Announce Plans for Synthetic Human Genomes," *Washington Post*, June 2, 2016.

4. Collins, quoted in ibid.

5. Charles Lieber, quoted in Simon Makin, "Injectable Brain Implants Talk to Single Neurons," *Scientific American* (March 1 2016), https://www.scientificamerican.com/article/injectable-brain-implants-talk-to-single-neurons.

6. Drew Endy and Laurie Zoloth, "Should We Synthesize a Human Genome?," open letter, May 10, 2016, https://dspace.mit.edu/bitstream/handle/1721.1/102449/ShouldWeGenome.pdf?sequence=1.

7. J. Craig Venter, quoted in Maggie Fox, "Synthetic Stripped-Down Bacterium Could Shed Light on Life's Mysteries," NBC News, March 24 2016, http://www.nbcnews.com/health/health-news/little-cell-stripped-down-life-form-n545081.

8. Kurzweil, *The Singularity Is Near*, 296.

9. Templeton Prize–winner Holmes Rolston III has talked about "three big bangs" that mark the most significant developments in the history of the universe. These three bangs are the start of the universe itself, the start of life, and the start of mind. The Singularity, should it happen, is something Kurzweil might choose to classify as a fourth big bang.

10. Views associated with Descartes are known as *Cartesian* after the Latin translation of his name—Cartesius.

11. A field known as *simulation theology* has arisen to explore the link between ideas like Kurzweil's and Christian theology.

12. Quoted in Michael Specter, "Rewriting the Code of Life," *New Yorker*, January 2, 2017, 36, http://www.newyorker.com/magazine/2017/01/02/rewriting-the-code-of-life.

Chapter 11

1. Crutzen and Stoermer, "The Anthropocene," 17.

2. Antonio Stoppani, "Corso di Geologica," in *Making the Geologic Now: Responses to Material Conditions of Contemporary Life*, ed. Elisabeth Ellsworth and Jamie Kruse, trans. Valeria Federighi and Étienne Turpin (New York: Punctum Books, 2013), 34–41.

3. Thomas C. Chamberlin, *Geology of Wisconsin: Survey of 1873–1879* (Madison, WI: Commissioners of Public Print, 1883).

4. This is the same Andy Revkin who suggested that humanity might get queasy at the prospect of climate engineering.

5. David Keith, *The Case for Climate Engineering* (Cambridge, MA: MIT Press, 2013), 173.

6. Crutzen and Schwägerl, "Living in the Anthropocene."

7. Rick Bass, quoted in Bogard, *The End of Night*.

Further Reading

Ackerman, Diane. *The Human Age: The World Shaped by Us*. New York: W. W. Norton, 2014.

Biello, David. *The Unnatural World: The Race to Remake Civilization in Earth's Newest Age*. New York: Scribner, 2016.

Bogard, Paul. *The End of Night: Searching for Natural Darkness in an Age of Artificial Light*. New York: Little, Brown and Company, 2013.

Crutzen, Paul, and Eugene Stoermer. "The Anthropocene." *Global Change Newsletter* 41 (May 2000): 17–18.

Drexler, K. E. *Engines of Creation: The Coming Ear of Nanotechnology*. New York: Anchor Books, 1986.

Drexler, K. E. "Molecular Engineering: An Approach to the Development of General Capabilities for Molecular Manipulation." *Proceedings of the National Academy of Sciences of the United States of America* 78 (9) (1981): 5275–5278.

Ellis, Erle. 2012. "The Planet of No Return: Human Resilience on an Artificial Earth." *The Breakthrough Journal* (Winter). https://thebreakthrough.org/index.php/journal/past-issues/issue-2/the-planet-of-no-return.

Feynman, Richard. "There's Plenty of Room at the Bottom." Lecture given at California Institute of Technology, December 29, 1959. *Caltech Engineering and Science* 23(5) (February 1960): 34. http://calteches.library.caltech.edu/1976/1/1960Bottom.pdf.

Gardiner, Stephen. *A Perfect Moral Storm: The Ethical Tragedy of Climate Change*. New York: Oxford University Press, 2011.

Kurzweil, Ray. *The Singularity Is Near: When Humans Transcend Biology*. New York: Penguin, 2006.

Lee, Keekok. *The Natural and the Artefactual: The Implications of Deep Science and Deep Technology for Environmental Philosophy*. Lanham, MD: Lexington Books, 1999.

Leopold, Aldo. *A Sand County Almanac: And Sketches Here and There*. New York: Oxford University Press, 1949.

Mark, Jason. *Satellites in the High Country: Searching for the Wild in the Age of Man*. Washington, DC: Island Press, 2015.

Marris, Emma. *Rambunctious Garden: Saving Nature in a Post-Wild World*. New York: Bloomsbury, 2011.

Marris, Emma, Peter Kareiva, Joseph Mascaro, and Erle Ellis. "Hope in the Age of Man." *New York Times*, December 7, 2011. http://www.nytimes.com/2011/12/08/opinion/the-age-of-man-is-not-a-disaster.html.

Marsh, George Perkins. *Man and Nature: Or, Physical Geography as Modified by Human Action*. Cambridge, MA: Harvard University Press, 1965 [1864].

McKibben, Bill. *The End of Nature*. New York: Random House, 1989.

McKibben, Bill. *Enough: Staying Human in an Engineered Age*. New York: Henry Holt and Company, 2003.

Mill, John Stuart. "On Nature." In *Nature, the Utility of Religion, and Theism*. London: Longmans, Green, Reader, and Dyer, 1874.

Morton, Oliver. *The Planet Remade: How Geoengineering Could Change the World*. London: Granta, 2015.

Pearce, Fred. *The New Wild: Why Invasive Species will be Nature's Salvation*. Boston: Beacon Press, 2015.

Purdy, Jedediah. *After Nature: A Politics for the Anthropocene*. Cambridge, MA: Harvard University Press, 2015.

Purdy, Jedediah. "The New Nature." *Boston Review*, January 11, 2016. http://bostonreview.net/forum/jedediah-purdy-new-nature.

Vince, Gaia. *Adventures in the Anthropocene: A Journey to the Heart of the Planet we Made*. Minneapolis, MN: Milkweed, 2014.

Index